Fifth Edition
GENERAL CHEMISTRY
101/102 Laboratory Manual

Department of Chemistry and Biochemistry
University of North Carolina Wilmington

Edited by Cecilia B. Kieber, Robert J. Kieber Jr. and Charles R. Ward

Kendall Hunt
publishing company

This lab manual is provided by the UNCW Department of Chemistry and Biochemistry. Royalties from this lab manual are returned to the Chemistry Department to fund the purchase of specialty software, glassware, and supplies used for developmental and instructional purposes for the General Chemistry laboratories.

Cover image © Shutterstock.com

Kendall Hunt
publishing company

www.kendallhunt.com
Send all inquiries to:
4050 Westmark Drive
Dubuque, IA 52004-1840

Copyright © 2002, 2004, 2007, 2009, 2012 by University of North Carolina Wilmington

ISBN 978-1-5249-3469-9

Published in the United States of America

Contents

Preface

The purpose of the chemistry laboratory experience is to enhance student learning through performing hands-on activities. First, the laboratory experiments will allow you to apply concepts studied in class. Many of the experiments in this manual have been designed to introduce new concepts and procedures that can only be learned through hands-on investigation of chemical substances and manipulating laboratory equipment. The general chemistry laboratory program is designed to teach common laboratory skills and safety practices which are the building blocks for more advanced lab classes, and for the workplace and research environment. Finally, the lab experience is a mechanism to stimulate inquisitiveness and induce the thrill of scientific discovery. Students in chemistry lab will learn that scientific experiments do not always follow prescribed conclusions, and may lead to more questions which are the basis for studying the exciting challenges in science.

How much you get out of your laboratory experience depends on how much you put into it. Unlike other courses failure to prepare in advance for lab can lead not only to a low grade but also to accident or injury to yourself or others as well. It is very important to read each experiment before coming to lab. Pay special attention to Safety Precautions for each experiment. Make notes of anything that you do not understand and ask questions about these problems during lab. Reading the manual and answering Prelab Questions before class allows you to be well prepared, conscientious, and able to work cooperatively as part of a team in order for you to learn as much as possible in this hands-on course.

After you have completed each experiment you will be asked to re-evaluate and apply the main concepts you learned by answering lab review questions. Your written communication skills will be applied towards the Calorimetry experiment, where you will be required to formalize a lab report on your own. This task allows you to describe the purpose of the experiment and use the information gained towards predicting and explaining some of the phenomena in the world you live in. Finally, as a method for improving student critical thinking skills, the General Chemistry curriculum includes one or more Laboratory Practicals per semester. Lab Practical activities allow each student to take charge in designing and performing their own experiment to answer a specific problem.

Thus, the main objective of your laboratory experience is not only to learn by hands-on manipulation, but to create an atmosphere conducive to improving inquisitiveness, application of critical thinking skills, and an excitement for scientific study.

Cecilia B. Kieber
Robert J. Kieber Jr.
Charles R. Ward

Safety Guidelines

In the General Chemistry Laboratory, safety is of paramount importance. Learning to handle potentially harmful chemicals in a safe manner is one of the skills you will learn in this course. To help ensure your safety, the following guidelines should be practiced at all times. Failure to do so will result in your expulsion from the laboratory. Please read these precautions carefully and then answer the questions on the safety contract that follow. Sign and date the contract and turn it in to your laboratory Instructor before proceeding with any experiment.

Eyes

In order to meet the requirements in your chemistry lab, **safety glasses are required to be worn over your eyes at all times in the laboratory during experimentation and clean-up of glassware.** The glasses must conform to the requirements of ANSI Z87.1-2003, indicated by the "Z87" designation inscribed on the frame or earpiece of the glasses. You must provide your own safety glasses or safety goggles, which can be purchased in the bookstore. If you wear prescription eyeglasses, the bookstore sells safety glasses that must be worn over them. Wearing contact lenses are not recommended in lab, and should be replaced by wearing prescription glasses.

Failure to bring safety glasses will result in your expulsion from lab. Students are not to remove safety glasses/ goggles until all students have completed the experiment. Your Instructor will notify the class when it is safe to remove safety glasses/goggles. Disregard of the safety glasses rule, or failure to heed your Instructor's enforcement of this policy, will result in your expulsion from lab.

Note and stay aware of the location of the nearest eyewash fountain in the event of a chemical splash on your face and around your eyes.

Ingestion

Regard all laboratory chemicals as potential hazards. Therefore, open containers of food or drink are not permitted in the laboratory as possible toxic materials can inadvertently contaminate them. However, you may step out into the hallway to eat or drink if necessary. Never eat, drink, or taste anything used in the chemistry laboratory. Do not use your mouth to suction liquids through a pipet. Disregard of these rules will result in a warning and a 5 point deduction; a second and further offense during the semester of ingestion in the lab will result in your expulsion from the lab.

Respiratory

Noxious or potentially hazardous vapors can be produced by some chemicals and by some chemical reactions. For your protection, these chemicals will usually be located in the fume hoods. Do not remove these chemicals from the fume hoods. When obtaining or disposing of chemicals in the fume hood, be certain that the sash (window) is low enough to cover your face and neck to prevent accidental inhalation of noxious materials. If you are asked to determine the odor of a certain chemical, you should fan the fumes gently towards your nose with your hand (called "wafting"), or pass the cap of the test tube quickly by your nose, without directly inhaling the fumes. Smoking is strictly prohibited in the laboratory.

Skin

Handle all chemicals in such a way as to avoid contact with your skin. Utilize proper laboratory equipment (spatulas, tongs, pipets, forceps, etc) for the transport and delivery of chemicals. Use crucible tongs or paper towels to handle hot objects. Be sensible concerning the clothing you wear in the laboratory in order to reduce exposing your skin to chemical contamination. Long hair must be tied back so that it cannot accidentally come in contact with the burner flame. Rubber bands may be provided if necessary for long hair. It is strongly recommended that close-toed shoes be worn in the laboratory to prevent spills and broken glass accidents on exposed skin. Your Instructor will remind you to wear closed-toed shoes particularly during experiments using strong acids and bases. The Safety Precautions for each experiment provide additional information you can use as a guide in determining the type of protective apparel and shoes that should be worn during specific lab experiments.

Note the location of the safety shower to be used for any major chemical spill that gets on you. Wash your hands thoroughly after chemical contact and before leaving the laboratory.

Backpacks

Keep your workspace uncluttered and clear of nonessential items. Packs and bags for books, jackets, umbrellas, etc. should be placed on the floor against the wall or in the far back of the room on the unused lab benches, chairs, or coat racks. One chair is provided in the laboratory particularly for use by those who are injured. Lab stools must remain close to the lab benches to provide a clear path in the aisles so that an emergency situation can be accessed quickly.

Handling Equipment

Glassware

Be careful handling laboratory glassware. Inspect all glassware before use to determine if there are any cracks or breaks. Replace glassware as necessary and deposit broken glass in the proper container. Do not attempt to force frozen glass joints, stoppers, or stopcocks. Care must be used when inserting a pipet into a pipet bulb by holding the pipet close to the end being inserted and gently rolling the end into the bulb so that the pipet bulb is very loose on the end of the pipet. Heated glassware must be cooled down on the bench top first, and then allowed to cool in a water bath if required.

Flame

Always pay attention to the location of any Bunsen burner flames as they can be very difficult to see. Keep a lit burner away from the edge of your lab bench. Never leave a lit burner unattended. Be certain that either you or one of your team members keeps an eye on your lit burner if the other is obtaining materials, etc. Extinguish the burner flame if it is not in use. Keep solvents, such as alcohol and acetone, away from the flame. When you are finished with your experiment, make certain that the gas valve at the lab bench is closed completely. Never cap a vessel being heated as the gases produced by the heat or reaction may build up enough pressure to cause an explosion. Mix chemicals in small quantities as larger quantities may cause violent reactions. Note the location of the fire extinguisher and learn how to use it. Know where the fire alarm is located in your lab as well as the location of the nearest telephone and exit.

Fire

Shout FIRE to alert your classmates and Instructor if there is a fire, no matter the size of the flames. If the fire is in a test tube or small container it can be contained with a watch glass or beaker turned upside down to smother the flames. If the flame cannot be extinguished by a fire extinguisher, sand, or water, the class will need to evacuate the building immediately. In the event that clothing catches on fire, the fire needs to be smothered by wrapping the student in a lab coat, fire blanket, or rolling the student on the floor.

Handling Chemicals

Handle all chemicals with care. Read the labels of the chemicals you are using and be certain that you are using the proper chemicals and the proper concentrations of the chemicals. Consider, for example, the similarities between the following chemical names: sodium nitrate and sodium nitrite. Both behave very differently from each other so it is essential that you know and obtain the correct chemical for your experiment. In addition, there may be two or more containers of the same reagent but with different concentrations such as 0.1 M and 1.0 M. Again, it is important that you are aware of what concentration of each reagent is required in order for your experiment to run predictably, smoothly, and safely.

Clean and dry glassware and equipment are to be used for obtaining chemicals. Do not remove reagent bottles from their original location. It is wiser to take your container to the front lab bench or side shelf to dispense chemicals. Do not insert your spatula or pipet directly into a reagent bottle. Pour a portion of the reagent into a beaker or graduated cylinder and take what you need from this container. It is always possible to obtain more material if necessary. Use only the minimum volume required from the reagent bottle and make certain to place the cap or stopper back on the correct bottle immediately. To avoid contamination, never pour excess or unused reagent back into the original reagent bottle. Label all chemicals and waste receptacles at your bench in order for you to be organized and aware of the identity of the materials at your workspace.

Material Safety Data Sheets (MSDS) are available for each chemical used in the laboratory and contain information concerning health hazards and reactivity. These are available in a notebook at the instructor's station in the laboratory. An example of a MSDS form for the chemical ethanol is located in **Appendix A**.

Obtain assistance from your instructor in case of a major spill. All chemical spills should be cleaned up immediately with water and wet paper towels. Keep your workspace clean and free of spills. Maintain the weighing balance you use by wiping up any chemical spills with paper towels.

Acids and Bases

Use caution when working with acids and bases as these are caustic and cause burns to the skin and will create holes in your clothes. Always add an acid slowly into water. Never add water into an acid because the heat evolved may cause the acid to splatter, which could result in serious injury.

Disposal of Chemicals

Dispose of all wastes properly according to the experimental procedure. Solid wastes should be thrown away in the appropriate garbage can if there are no plans for recycling. If the procedure specifies to pour aqueous, non-toxic waste solutions down the sink at your bench, flush the drain with plenty of water.

Accidents and First Aid

Report every accident or injury to your Instructor immediately in order to expedite the appropriate first aid. An Accident Report Form must be completed by your Instructor for all accidents requiring any form of first aid.

Eyes

Immediately flush the contaminated eye using the eyewash or the tap at your sink for 15 minutes.

Burns

All heat burns should be flushed thoroughly with plenty of cold, running water for at least 10 minutes in the nearby sink, hose, or under the shower. An ice pack may also be applied. Act quickly. Strong chemicals can burn the skin rapidly; therefore flush the area with copious amounts of water immediately by shower, hose, or nearby faucet. Remove and replace clothing, socks, and shoes that have been affected. Do not attempt to neutralize chemical burns as the heat of the reaction can cause further injury.

Cuts

All cuts should be treated as though chemicals may be present and flushed well with water, washed with soap and water, and flushed again. Obtain bandages from your instructor if necessary. If you suspect glass fragments are present, inform your Instructor and visit the University Health Center promptly. An Accident Report Form must be completed by your instructor for all cuts received in the laboratory.

Preparedness

In order for you to be continuously aware of the safety notes, procedures, and waste disposal methods for each experiment you perform in this course, you are required to have the current edition of the General Chemistry 101/102 Laboratory Manual. You are the person most responsible for your own safety. Therefore, it is essential to **read the laboratory procedure carefully before beginning your work, noting any safety precautions**. Learn what is expected of you during the lab period and come to class with any questions you may have concerning the theory or procedure.

Laboratory Policy and Grading

Safety Glasses/ Goggles

Students must have safety glasses or safety goggles to perform an experiment. Students are required to purchase and bring their own safety glasses to lab. The consequence for not having safety glasses/goggles will be that a student cannot participate in the experiment and will earn a 0 for that lab grade.

Lab Manual

Students are not permitted to attend the class without the current edition of the UNCW General Chemistry 101/102 Laboratory Manual. Copies of the experiment are not permitted beyond the first week of class (these will be provided only the first week by the department).

Lab Course and Credits

Lab is part of the 4 credit General Chemistry course, CHM 101 or CHM 102. Students must attend the 3-hour laboratory class in order to get all 4 credits, even if the class is being repeated. Chemistry is an experimental science; therefore lab is an integral part of the entire course.

Absenteeism and Make-Ups

All absences will result in a grade of 0, *no matter the reason* for the absence. Lab and lab practical make-ups are *not allowed regardless of reason*. Lab work is permitted only during scheduled lab periods. Students are only allowed in lab while under the supervision of their Instructor. Unauthorized experiments on weekends or after class hours are not permitted. The lowest grade received from lab and the lowest grade on one of their quizzes will be dropped in tabulating the course grade. Lab Practicals and formal laboratory reports are mandatory activities; thus, the grade earned from these activities will be tabulated in the final lab grade. Students who miss more than 4 laboratory sessions will receive a grade of "F" for the entire CHM course. As a courtesy, the lab Instructor will warn the student and notify the professor of the course in the event the student has missed 3 labs.

Preparation

Department policy dictates that students entering the classroom 10 minutes or more and have missed most or all of the pre-lab briefing, will not able to participate in that laboratory class. This is to ensure students have the opportunity to hear the safety precautions and laboratory expectations regarding the experiment.

Laboratory Logistics

The lab course is set up into the following major sections:

1. Pre-lab briefing including power point presentation, discussion of safety, and demonstration of lab techniques. Most lab experiments will require a short 15-20 minute briefing and demonstration on the procedure, technique and/or theory. Key points and concepts regarding the experiment and safety will be discussed. Focus will be on teaching proper scientific techniques.

2. Performing experiment, cleanup and data write-up. The experimental portion of the lab class should take approximately 1.5 to 2 hours. Students will write up data and conclusions of the experiment. Students will clean-up and maintain the glassware in their drawers. **Safety glasses/goggles must be kept on until all experimentation and cleaning is complete by all students.** Students who have finished this section of the lab class early may take a short break outside of class and return in 10-15 minutes when all students are finished.

3. Lab as recitation to lecture class. When all experimentation and cleanup has been completed, supplemental questions which relate to theory and concepts learned in lecture class will be assigned and discussed. This is a period of lab class in which all students may take off their safety glasses/goggles for the remaining 30-45 minutes of lab.

Pre-Lab Questions

Students must have all Pre-lab questions completed before the lab briefing and will not be accepted late. Pre-lab questions are located in the Lab Manual after each lab experiment. These questions encourage students to come prepared to lab by reading theory, operation, and safety of an experiment. The Instructor will be available 10 minutes before class to answer any questions regarding Pre-lab questions or theory of the experiment. Pre-labs will be graded 5 points for effort, and will be handed in with all paperwork (data sheets, supplemental questions, etc) to be graded a second time for correctness for an additional 5 points.

Lab Drawer and Equipment

Students will work in pairs and will use a lab drawer appropriated for that particular time slot (for example: CHML 101 which meets Mon 11-2 will use the lab drawer labeled '101 Monday'). Proper spacing will allow 2 pairs of 4 students per lab row so that each pair has access to a sink and gas outlet. Students will not be allowed to remove glassware and equipment from other bench drawers. Missing lab equipment can be obtained from the front cabinets which contain extra equipment and glassware. Students will be graded weekly on maintenance and cleanliness of lab equipment in their lab drawers. The cabinets under each sink are designated to contain 1 or 2 ring stands. Extra ring stands are located on bookshelves in each lab. Students should be responsible for returning ring stands and equipment to proper locations. Aisles in the lab should be clear of backpacks, jackets, and skate boards, etc. so students can move freely and avoid spills, etc.

Check-In/Check-Out of Lab Drawers

Students will refer to the Chemical Equipment Checklist in the Laboratory Manual directly following the Safety Guidelines to check-in to their lab drawer. This sheet will be used as a reference guide throughout the semester. Extra equipment and glassware are kept in the cabinets below the whiteboards in each lab. If the drawer is dirty, students should clean and dry the drawer before lining it with paper towels and restocking with equipment. At the end of the semester, students will check-out their drawers and restock glassware and equipment using this Chemical Equipment Checklist. Lab Instructors will be responsible for checking over this sheet and signing students out of lab, awarding a lab maintenance grade of 5 points.

Special Equipment and Chemicals

All chemicals reagents and special equipment needed for a particular experiment will be located on the front or side reagent bench and on carts at the side or front of the room. These locations will be conveyed to students by the laboratory Instructor. All specialized glassware and equipment must be returned to their original location, cleaned well with DI water and dried. Students must be careful not to contaminate stock chemical and reagents with spatulas or pipets. It is recommended that students obtain small amounts of stock solutions and chemicals by placing in a cleaned and dried beaker, test tube or flask from their lab drawer to bring the materials back to their bench. The student can always obtain more chemical if this is not sufficient. Students must not pour or return unused chemicals back into original stock containers to prevent possible contamination. Students must not use spatulas/scoopulas other than those cleaned and dried from their own lab drawer to obtain chemical solid materials. Notify your Instructor if glassware has been broken. Broken glassware will be placed in the broken glass container in the lab.

Lab Grade

The laboratory portion of the CHM 101 and 102 course will be worth 15% of the overall grade. Each lab is worth 45 points. The lowest lab grade will be dropped from the overall lab course grade, with the exception of laboratory practicals and formal lab reports, which cannot be dropped from the final grade.

The following describes the individual point allocation for each lab experiment:

Pre-lab Questions = 5 points for effort/ 5 points for correctness

Lab Technique and Maintenance/ Abiding by Safety Rules/ Attitude and Preparedness = 5 points

Data Sheet and Questions = 20 points

Supplemental Lecture Questions = 10 points

Safety and Lab Policy Contract CHML 101

1. What is the best way to determine the odor of a chemical?

2. Designate the location of the following items in this room?
   a) Fire extinguisher
   b) Fire alarm
   c) Safety shower
   d) Eyewash fountain

3. Describe some of the dangers of not wearing safety glasses in the laboratory.

4. What should you immediately do if chemicals come in contact with your skin?

5. What is the best way to be prepared for the laboratory experiment in order to work efficiently and effectively?

6. Describe the potential problems of bringing an opened bottle of Pepsi into the lab.

7. What does "MSDS" stand for? Using the MSDS information provided as an example in **Appendix A**, write the density and boiling point of ethanol in the space below. What are the potential health effects of ingesting ethanol?

I have read, understand, and will follow the UNCW General Chemistry 101/102 Laboratory Manual Safety Guidelines and Laboratory Policy. I understand that failure to comply with the Safety Guidelines/Lab Policy will result in point deductions and/or expulsion from lab:

Name (Printed): _____ **Section:** _____

Signature: _____ **Date:** _____

Safety and Lab Policy Contract CHML 102

1. Why is it necessary to check reagent names and their concentrations before use?

2. What is the proper height or placement of the sash during use of the fume hood?

3. What are the potential problems of placing a pipet bulb onto the end of a pipet? Describe the proper method for inserting a pipet into a pipet bulb.

4. Where is the appropriate place to pour excess reagents?

5. Why should you label reagents at your bench?

6. Describe the proper method for obtaining reagent chemicals from the front lab bench.

7. Refer to the MSDS example (**Appendix A**) to describe the information provided about the epidemiology of ethanol. What are some conditions to avoid in terms of ethanol's stability and reactivity?

I have read, understand, and will follow the UNCW General Chemistry 101/102 Laboratory Manual Safety Guidelines and Laboratory Policy. I understand that failure to comply with the Safety Guidelines/Lab Policy will result in point deductions and/or expulsion from lab:

Name (Printed): _____ **Section:** _____

Signature: _____ **Date:** _____

Common Laboratory Glassware and Equipment

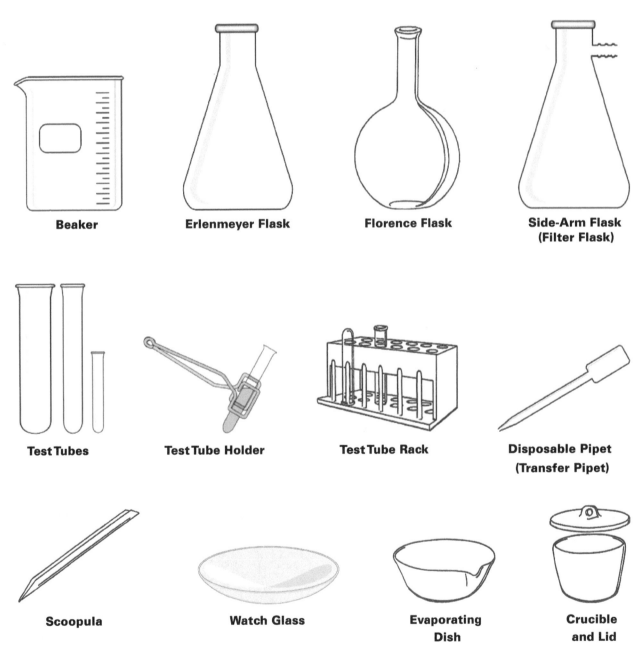

Beaker

Erlenmeyer Flask

Florence Flask

Side-Arm Flask
(Filter Flask)

Test Tubes

Test Tube Holder

Test Tube Rack

Disposable Pipet
(Transfer Pipet)

Scoopula

Watch Glass

Evaporating
Dish

Crucible
and Lid

Short-Stem Funnel

Büchner Funnel

Wash Bottle

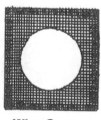

Wire Gauze

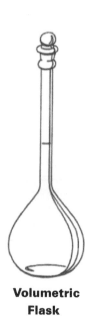

Volumetric Flask

Thermometer

Graduated Cylinder

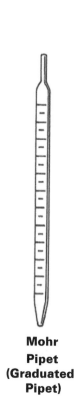

Mohr Pipet (Graduated Pipet)

Volumetric Pipet

Pinchcock Clamp

Crucible Tongs

Mortar and Pestle

Hoffman Screw Clamp

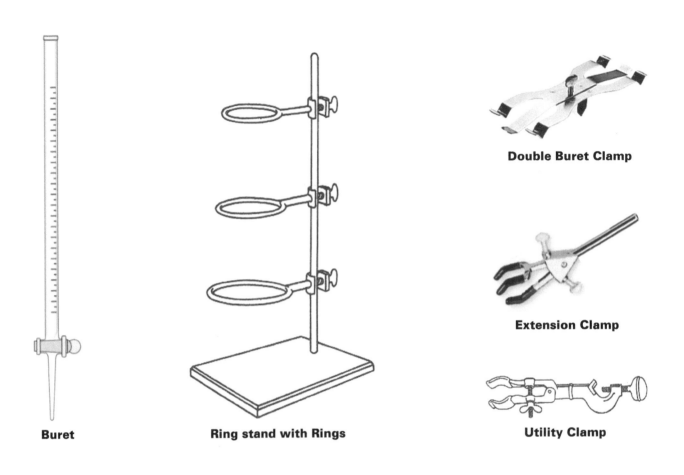

Buret

Ring stand with Rings

Double Buret Clamp

Extension Clamp

Utility Clamp

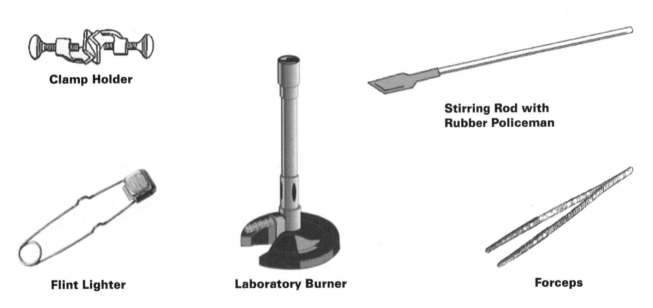

Clamp Holder

Flint Lighter

Laboratory Burner

Stirring Rod with Rubber Policeman

Forceps

(If the burner has a control for both gas and air, it is a **Tirrill burner.** If there is only an air control, it is a **Bunsen burner.** However, it is common to refer to all laboratory burners of this type as Bunsen burners.)

General Chemistry Equipment Checklist 101

(Leave in Lab Manual for reference until end of semester)

Lab Drawer contents:

Beakers:

_____ (1) 600 mL

_____ (1) 400 mL

_____ (1) 250 mL

_____ (1) 100 mL or 150 mL

_____ (1) 50 mL

Test tubes:

_____ (10) standard test tubes

_____ (1) large test tube

_____ (4) micro test tubes

_____ (1) test tube holder

_____ (1) test tube rack

Erlenmeyer flasks:

_____ (2) 250 mL

_____ (1) 125 mL

_____ (1) 50 mL

Graduated cylinders:

_____ (1) 10 mL

_____ (1) 25 mL or (1) 50 mL

Miscellaneous Equipment:

_____ (1) plastic DI water squeeze bottle

_____ (1) funnel

_____ (1) spatula/scoopula

_____ (1) glass stirring rod (with or without rubber policeman)

_____ (1) watch glass

_____ (1) forceps

_____ (1) flint lighter

Lab Drawer maintenance is the responsibility of the student. Periodic inspections by Instructor will result in points gained or deducted depending on proper maintenance and cleanliness of glassware and equipment.

Name (printed) _____ **Lab Drawer** _____

Signature _____ **Section** _____

Instructor's initials for check-out at end of course _____

General Chemistry Equipment Checklist 102

(Leave in Lab Manual for reference until end of semester)

Lab Drawer contents:

Beakers:

_____ (1) 600 mL

_____ (1) 400 mL

_____ (1) 250 mL

_____ (1) 100 mL or 150 mL

_____ (1) 50 mL

Test tubes:

_____ (10) standard test tubes

_____ (1) large test tube

_____ (4) micro test tubes

_____ (1) test tube holder

_____ (1) test tube rack

Erlenmeyer flasks:

_____ (2) 250 mL

_____ (1) 125 mL

_____ (1) 50 mL

Graduated cylinders:

_____ (1) 10 mL

_____ (1) 25 mL or (1) 50 mL

Miscellaneous Equipment:

_____ (1) plastic DI water squeeze bottle

_____ (1) funnel

_____ (1) spatula/scoopula

_____ (1) glass stirring rod (with or without rubber policeman)

_____ (1) watch glass

_____ (1) forceps

_____ (1) flint lighter

Lab Drawer maintenance is the responsibility of the student. Periodic inspections by Instructor will result in points gained or deducted depending on proper maintenance and cleanliness of glassware and equipment.

Name (printed) _____ Lab Drawer _____

Signature _____ Section _____

Instructor's initials for check-out at end of course _____

Lab Practical Information

Introduction

The purpose of most general chemistry laboratory courses is to develop analytical techniques and reinforce concepts learned in lecture class. Following detailed procedures to reach prescribed conclusions will successfully fulfill the requirements for the majority of experiments in these courses. However, the major shortcoming of this type of laboratory instruction is that it fails to develop problem solving and advanced reasoning skills, or to encourage inquisitiveness. UNCW Department of Chemistry and Biochemistry has addressed this deficiency by incorporating laboratory practicals throughout the entire general chemistry curriculum. Laboratory practicals are designed to encourage critical thinking by posing a new question which is similar to experiments performed in lab, but which present a conceptual "twist". Our main objectives in instituting lab practicals into the general chemistry curriculum include the following:

1. Reinforce concepts, skills, and techniques learned in lab and lecture class
2. Encourage creative and independent thinking
3. Improve problem solving skills
4. Provide a unique discovery-based learning experience where students are able to develop a procedure, perform their own experiment, and interpret the data collected
5. Increase excitement and confidence in scientific experimentation

Students are able to design and perform their own experiment, and incorporate data, observations and calculations into their conclusion to successfully answer the practical problem. Chemistry concepts learned in lab and lecture can be applied to solve the problem. Students should have a good understanding of the techniques and skills involved in performing the experiments learned, and the necessary glassware and equipment for that experiment.

Grading includes demonstration of lab skills and abiding by safety considerations. Emphasis in evaluating practicals remains on the actual problem solving strategies and thorough understanding the underlying scientific concepts of the experiment.

General Information About Lab Practicals

The concept and logistics of laboratory practicals will be introduced during the first week of lab and includes a power point presentation. Examples of a practical activity, student worksheet and evaluation sheet are provided on the bulletin board located outside the chemistry labs and on the general chemistry laboratory website.

During the normal academic year, lab practicals will be administered twice in one semester: once mid-semester, and the second during the final week of the semester. Practicals cover only the lab and lecture material learned up to that point, and are not cumulative (with the exception of the summer chemistry courses). Lab practicals are an individual effort; therefore students should be discouraged from relying on the efforts of their partner during lab.

Lab practical experiments are learning activities and students can be aided throughout the practical if necessary by asking questions of their Instructor and receiving some hints (points are not deducted for the first question).

The class will be divided alphabetically into two groups of 12 each. Each group of 12 will attend an hour and twenty minute practical session, with students working an assigned lab station. No extra time will be allotted to finish the practical activity as sessions must run back-to-back all day. Instructors will list and email class time assignments a week prior to the practical.

Students will be required to perform only one lab practical activity during a lab practical session. There are numerous lab practical activities and each student will have his or her unique problem to solve. Lab practicals will be in place at each lab station when the students enter the lab. Each practical includes a title, the problem statement, and a list of available materials and possibly some formulas. A Student Worksheet will be assessed by the lab Instructor for all written work and thought processes. As an introduction to this program, practical activities administered in the first half of the CHML 101 semester will include a title which relates directly to the experiment performed in lab.

Each practical activity contains a list of available materials and formulas; however, some of these are not necessary to answer the problem. Students need to think about the correct glassware and equipment necessary to complete their experiment. Students will be able to use common glassware and equipment contained in the lab drawer nearest their station. Chemicals and specialized glassware will be located on the front and side reagent benches in the lab, and will be identified by an index card.

Instructors will be responsible for signing off on students' written procedures for the purpose of ensuring safety of procedural method and usage of appropriate quantity of materials ONLY. Instructors may not give information to correct the procedure during sign off. Students will be given the opportunity to modify procedures as the experiment progresses to supplement their first attempt. Students can still receive credit for realizing the way to solve the problem even if their original procedure does not lead them to the correct answer.

General Rules When Taking Lab Practicals

- Use of cell phones or personal computers is not permitted.

- Books and notes are not permitted.

- Calculators are not permitted during the writing and modifying of procedures. Calculators will be permitted upon discretion of instructor, or during data/observations/calculations and conclusions sections.

- Safety glasses/goggles must be worn while performing experiment and during clean-up of materials. 5 points will be deducted for borrowing a pair of goggles.

- Food and drink are not permitted unless in closed containers kept in personal belongings. Students may step into the hallway if drinking and eating is necessary during the practical session.

- Students must stay at assigned station. Students are not permitted to switch stations at any time.

- Talking is not permitted unless to Instructor.

Grading

Students will be graded on the scientific soundness, logic and clarity of their procedure and their experimental design, the organization of their data, and the explanations used in discussing conclusions. Lab practicals are worth 45 points each, the same points awarded each weekly lab experiment. Participation in the lab practical activity is mandatory; therefore, the grade earned from the practical may not be dropped from the overall lab grade.

A grade of 0 will be awarded to a student who does not attend his/her practical session. A grade of 0 will be awarded to a student who participates in the practical, but makes no attempt to solve the problem, or whose attempt does not match the activity. A grade of 20 will be awarded to a student who makes an effort to solve the practical problem, but whose answered problem is not scientifically sound. Practicals will be returned to students only temporarily to observe performance and will be collected by Instructor. A penalty of replacing original grade with a 0 will result in lab practicals which are not returned.

Aspirin Synthesis

Introduction

Organic chemistry originally began as the study and reactions of compounds formed by living organisms. Organic compounds are predominantly composed of carbon and are routinely synthesized in laboratories today. **Aspirin** is an organic compound called **acetylsalicylic acid**, and is one of the oldest and most common of all synthetic pain relievers. Aspirin has the structural formula shown in Figure 1 below.

Many organic compounds contain a **benzene ring**, which is a six-membered carbon ring containing three double bonds. Benzene is present in the central portion of the Aspirin molecule. The structure of benzene is usually abbreviated as C_6H_6, or as shown in Figure 2 below.

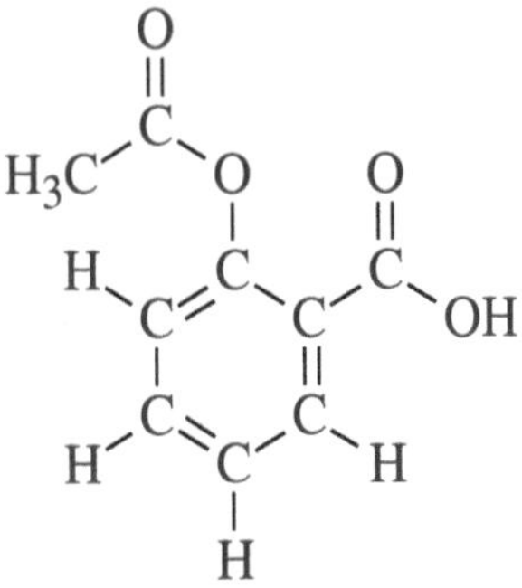

FIGURE 1 — *Structure of Aspirin*

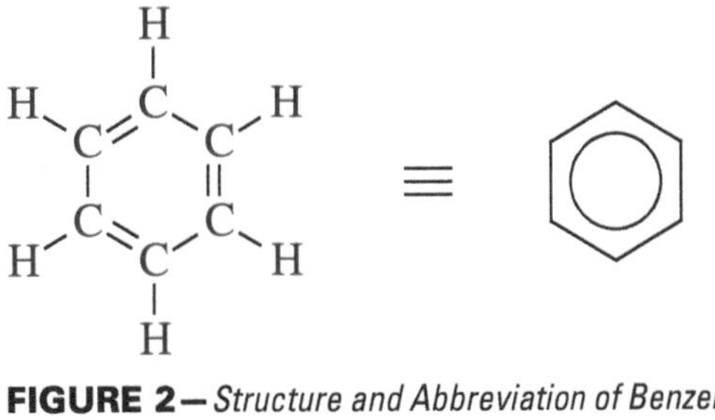

FIGURE 2 — *Structure and Abbreviation of Benzene (C_6H_6)*

Other painkillers have the benzene ring as the central portion of their structure. Examples of common painkillers include benzocaine (found in Bactine), methyl salicylate (oil of wintergreen), and eugenol (oil of cloves).

Many native cultures have long realized the curative powers of natural sources such as willow tree bark and meadowsweet flower. In 1763, the willow bark's power for "curing the agues" was finally put into print. In 1853, the compound **salicylic acid** was isolated from both willow bark and meadowsweet flower and now can be made in the laboratory.

Salicylic acid was used therapeutically as an analgesic (pain reliever), antipyretic (fever reducer), and anti-inflammatory (reduce swelling) drug. Although effective, salicylic acid, due to its high acidity, produced undesirable side effects such as irritation to the membranes of the mouth, throat, and stomach. A German chemist working for Friedrich Baeyer and Company synthesized a derivative of salicylic acid in 1893 called **acetylsalicylic acid,** which is less acidic. Baeyer began to market the drug under the trade name *Aspirin*, a name that gives credit to the acetyl group ("A") and to the Latin name *Spiraea ulmaria* ("spir") of the meadowsweet flower in which salicylic acid was first extracted.

Derivatizing salicylic acid involved a reaction called an **esterification reaction,** in which the hydrogen atom in the hydroxyl (–OH) group of salicylic acid is replaced with an acetyl group (–COCH$_3$). The acetyl group weakens the acidity of the carboxylic acid group (–COOH). The second reactant in synthesizing Aspirin, **acetic anhydride,** is effective at donating an acetyl group. **Phosphoric acid** (H_3PO_4) serves as a **catalyst** to speed up the reaction by supplying hydrogen ions to ionize acetic anhydride. A catalyst is any substance that speeds up a chemical reaction but is not used up in the reaction. The second product from the reaction is acetic acid, which is the main ingredient in vinegar. The chemical reaction for the synthesis of Aspirin is shown in Figure 3 below.

Even though the acetyl group masks the acidity of salicylic acid, the human body slowly decomposes acetylsalicylic acid back to salicylic acid (the therapeutic compound) in the bloodstream. Because of this, Aspirin has the tendency to irritate the stomach lining and may cause loss of approximately 0.5 mL of blood for each 500 mg tablet ingested.

The biochemical action of Aspirin is not completely understood, but is known to halt the production of prostaglandins, which are involved in the body's immune responses. Prostaglandins are synthesized upon introduction of foreign substances into the body and are responsible for many physiological processes such as pain, fever, and local inflammation. By preventing prostaglandin synthesis in the body, Aspirin alleviates the symptoms of an immune response but is not a cure for the condition causing the immune response.

The goal of this laboratory experiment is to provide you experience in the preparation of an organic compound. You will synthesize Aspirin by reacting acetic anhydride with salicylic acid as shown in the following equation in Figure 3 below. (Note that the coefficients of the reactants and products are all 1).

After you synthesize Aspirin, you are to determine the percent yield of your product. In order to calculate percent yield, it is necessary to determine the theoretical yield, which is dependent on the limiting reactant.

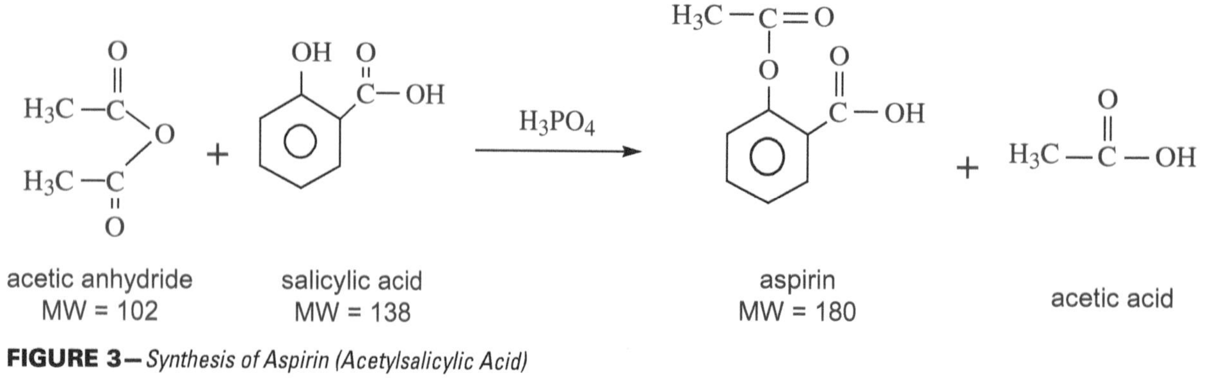

acetic anhydride　　　salicylic acid　　　　　　　　　　　　　aspirin　　　　　　　　acetic acid
MW = 102　　　　　　　MW = 138　　　　　　　　　　　　　　MW = 180

FIGURE 3 — *Synthesis of Aspirin (Acetylsalicylic Acid)*

Procedure

Using your own clean and dry scoopula/spatula, weigh 2.0 g of salicylic acid on a piece of weighing paper (recognized by its square shape and light consistency; do not confuse weighing paper with the round and thick Whatman filter paper). Measure 5.0 mL of acetic anhydride (located in the fume hood) in a clean graduated cylinder. Combine the salicylic acid and the acetic anhydride in a clean and dry 50 mL Erlenmeyer flask. Add 5 drops (approximately 0.25 mL) of 85% phosphoric acid (located in the fume hood) to the flask. Stir the mixture with a glass stirring rod. **Caution:** Acetic anhydride and phosphoric acid can produce bad chemical burns. Rinse with plenty of water if you spill any on yourself. *Immediately clean up any drips with a wet paper towel.*

Clamp the flask to a ring stand and immerse it in a 400 mL beaker containing approximately 250 mL of water (see Figure 4 below). Heat the water bath to 75°C and maintain this temperature for 5 minutes. Stir the mixture in the flask every minute during the heating process. All of the solid in the flask should be dissolved. Stop heating and *cautiously* add 2 mL of DI water to the flask. Stir the mixture and remove it from the water bath. Cool the flask and its contents to room temperature.

Once the flask is cooled to room temperature, add 20 mL of deionized water while stirring. Place the flask in an ice bath to speed the crystallization process. If crystals have not formed after five minutes, vigorously scratch the inside of the flask with a glass rod. This produces a rough surface on which the crystals may form. Allow the flask to remain in the ice bath for five minutes after crystals first appear.

Collect the Aspirin crystals by vacuum filtration using a Büchner funnel (see Figure 5). Secure the side-arm flask by clamping it to a ring stand to prevent spillage. Place a piece of round Whatman filter paper in the Büchner funnel and moisten it with DI water until it is flush with the bottom of the funnel. Turn on the suction and then pour the mixture containing the aspirin crystals into the Büchner funnel. Once all the water has gone through the funnel, wash the crystals with a 5 mL portion of ice-cold DI water. Repeat the washing with another 5 mL portion of ice-cold DI water. Allow air to be drawn over the crystals by suction for three to five minutes until the crystals are dry and free of water. Periodically fluff the crystals with a spatula to hasten drying. When the Aspirin crystals appear to be dry, transfer them to your watch glass and weigh them. Record the result on the data sheet as **actual yield** (weight of Aspirin produced). This crude product may contain water and some unreacted salicylic acid. Therefore, if your product is not dry, your actual yield may be higher than the theoretical yield.

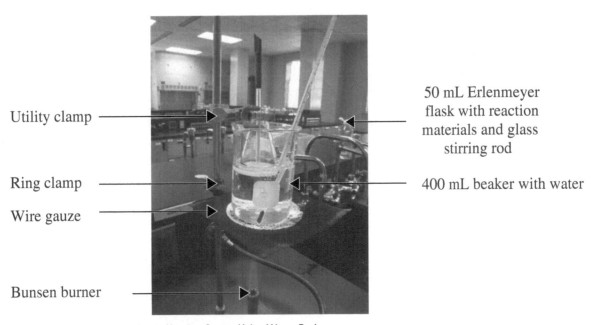

Utility clamp

Ring clamp

Wire gauze

Bunsen burner

50 mL Erlenmeyer flask with reaction materials and glass stirring rod

400 mL beaker with water

FIGURE 4 — *Aspirin Synthesis Heating Set-up Using Water Bath*

Utility clamp
Secured to
Ring Stand

Hose attached
to aspirator

Büchner funnel

Side arm flask

FIGURE 5 — *Filtration of Aspirin Crystals*

Calculations

The amount of product you could produce, the **theoretical yield**, is based on the amounts of starting materials used. You need to determine how much Aspirin you could theoretically produce using the amounts of each reactant (salicylic acid and acetic anhydride) and the **stoichiometric factor** for the reaction. One of the reactants will be the **limiting reactant** and will eventually run out, limiting the amount of Aspirin that can be produced. Follow the general formula below for each reactant used in the reaction (using the amounts available for each). You must compare the moles of product formed from each reactant to determine which reactant runs out and will form the least amount of product. In this experiment you will need to perform this calculation for the acetic anhydride reactant and the salicylic reactant, using the initial amounts provided for each.

Amount of reactant × mw of reactant × mole ratio of reactant to product = moles of product
(convert to grams)   (g/mole)   (using stochiometry of reaction)

Once you have determined the moles of product that can be formed based on the available amount of limiting reactant, you can convert this value from moles to grams using the molecular weight of the product, using the following formula. The molecular weight of Aspirin, 180 g/mole, is used to calculate its theoretical yield.

Moles of product × mw of product = grams of product (theoretical yield)
(g/mole)

To calculate the **percent yield** of Aspirin use your actual yield from today's experiment and place this value (in grams) the numerator of the equation below. The theoretical yield of Aspirin (in grams) is used in the denominator.

$$\% \text{ yield} = \frac{\text{actual yield}}{\text{theoretical yield}} \times 100\%$$

Safety Precautions

Salicylic acid is an irritant; therefore it should be handled with care. Prevent eye, skin and clothing contact with acetic anhydride, which is a strong corrosive agent and a **lachrymator** (causes the eyes to become irritated and shed tears). When working with acetic anhydride, use the fume hood and avoid inhaling any vapors. Phosphoric acid should also be handled with care as it is a fairly strong acid and will cause burns of skin, eyes and clothes. H_3PO_4 is also kept in the fume hood.

Waste Management

Wash the Aspirin crystals down the sink with plenty of water or deposit them in the solid waste container (the yellow waste container). ***Do not ingest the Aspirin you made!***

Aspirin Synthesis Prelab Questions

Name _____ **Section** _____

1. a) Describe why it is necessary to cool down the reaction mixture slowly after heating for 5 minutes? What is this process and why is it happening?

   b) What would happen if the cooling occurred too quickly?

2. What is the role of a catalyst in a reaction? What is the catalyst in the Aspirin synthesis?

3. What are the by-products of the reaction mixture after filtration? Could any of these by-products produce an odor, and if so, which ones?

Aspirin Synthesis Data Sheet

Name _____ **Section** _____

Name of Lab Partner(s) _____

1. Mass of salicylic acid   _____

2. Volume of acetic anhydride   _____

3. Mass of acetic anhydride
   (density of acetic anhydride = 1.08 g/mL)   _____

4. Mass of Aspirin produced   _____

5. Theoretical yield of Aspirin   _____

6. Percent yield of Aspirin   _____

Calculations

1. Limiting Reactant Determination:

   a) Moles of salicylic acid reactant used in experiment

   b) Moles of acetic anhydride reactant used in experiment

   c) Which reactant is the limiting reactant?

2. Theoretical Yield (grams of Aspirin product which could be produced using the limiting reactant determined above):

3. Percent Yield of your Aspirin product:

Aspirin Synthesis Review Questions

Name _____ Section _____

1. Think of some reasons why a student may obtain a larger actual yield than the calculated theoretical yield of aspirin.

2. a) If a student were to use 5.0 g of salicylic acid and an excess of acetic anhydride in the synthesis of Aspirin, calculate the theoretical yield of acetylsalicylic acid in moles.

    b) Calculate the theoretical yield of acetylsalicylic acid in grams.

3. If the actual yield of Aspirin from the reaction in the above question is 3.50 g, what is the percent yield?

4. The following reaction was carried out in the lab:

    $$Na_2CO_3 + 2\,HCl \longrightarrow 2\,NaCl + H_2CO_3$$

    If 3.50 g of Na_2CO_3 (mw = 106 g/mol) are reacted with 1.85 g of HCl (mw = 36.5 g/mole), what is the limiting reactant in the reaction? What is the theoretical yield possible for this reaction?

Aspirin Synthesis Supplemental Questions

Name _____ **Section** _____

Atomic Emission

Introduction

The wavelengths of light emitted by an excited atom are characteristic of the atoms of a particular element and can be used to identify the element. For example, sodium and potassium in blood serum and urine are routinely analyzed in a clinical laboratory by **atomic emission spectroscopy**. In addition, researchers are able to determine the chemical composition and age of cellestial bodies such as stars by analyzing the wavelengths of the radiation they emit.

Radiant light of wavelenths in the visible spectrum is emitted by hot atoms whose electrons have been promoted to excited states by a flame. German physicist and Nobel Prize winner (1918) Max Planck theorized that when a hot object emits light, there is a minimum quantity of energy that is associated with the emitted radiation. The minimum quantity of energy is called a **quantum**. Planck's ideas formed the basis for **quantum theory**. One of the basic principles of quantum theory is that atoms or molecules can absorb and emit only certain amounts of energy and these energies are associated with electrons occupying energy levels, or orbitals, in atoms or molecules.

Consider a sodium atom (Na) which is in the **ground state**, the lowest energy state of an atom. There are two electrons in each of the *1s* and *2s* levels, six in the *2p* level, and one electron in the *3s* level. The electron of interest in chemical reactions is the valence electron in the *3s* level because this electron is farthest from the nucleus and held less tightly by the protons in the nucleus. When a sodium atom is heated by a flame, the electron in the *3s* level absorbs energy and is transferred to a higher energy level. The sodium atom is now in an **excited state**.

When the excited electron of a sodium atom returns to its ground state energy level, it will emit a **photon** (a quantum of electromagnetic radiation) of the same energy as was absorbed. For sodium, this photon of light has a specific wavelength of 589 nm (yellow light). Planck realized that as an electron moves from one energy level to another in an atom, a precise amount of energy must be absorbed or emitted by the atom. Therefore, the difference in energy between any two energy levels in an atom can be calculated by measuring the **wavelength**, λ, of the photon emitted by the atom.

For example, **flame atomic emission** has provided important data supporting the idea of quantized energy. The flame of the Bunsen burner is hot enough to produce excited atoms and therefore can be used for flame emission. Electrons from methane and oxygen combustion reactions in the flame reduce sodium ions to sodium atoms (1). Sodium atoms are promoted to an excited state due to heat energy from the flame (2).

(1) $Na^+_{(g)} + e^- \longrightarrow Na_{(g)}$

(2) $Na_{(g)} \longrightarrow Na^*_{(g)}$ (excited state)

The electrons in the excited sodium atoms each emit a photon of light of a characteristic wavelength, λ, when they return to a lower energy state (3). In the last stage, sodium atoms are converted to sodium ions after gaining sufficient energy from the flame (4).

(3) $Na^*_{(g)}$ (excited state) $\longrightarrow Na_{(g)}$ + **photon (λ)**

(4) $Na_{(g)} \longrightarrow Na^+_{(g)} + e^-$

The energy, **E**, of the emitted photon can be determined using the following equation.

$$E = h \times \frac{c}{\lambda}$$

In the equation, **h** is **Planck's constant** (6.626×10^{-34} J ·sec), **c** is the velocity of electromagnetic radiation (commonly called the "speed of light", 2.998×10^8 m/s), and λ is the **wavelength** of the emitted photon. Note that the *energy of a photon is inversely proportional to its wavelength*. Therefore, a photon of red light (650 nm) has less energy than a photon of blue light (460 nm).

The **frequency** of electromagnetic radiation, υ, is the number of complete oscillations that an electromagnetic wave makes each second. Frequency is often expressed in several different units including reciprocal seconds (1/sec), s^{-1}, cycles/second (cps), and hertz (Hz). Although hertz (Hz) is the standard unit for frequency, it is necessary to use s^{-1} as the unit of frequency when performing calculations. The frequency of electromagnetic radiation, υ, is related to its wavelength, λ, by the following equation:

$$\upsilon = \frac{c}{\lambda}$$

Thus, the energy of the photon emitted from an excited atom can also be calculated knowing the frequency of the light emitted:

$$E = h\upsilon$$

In this lab exercise, you will use a **Boreal spectroscope** to observe and measure the wavelength of light emitted by gases in emission tubes. You will then chose one wavelength from the several spectral lines you observed and calculate the energy per photon and energy per mole of photons of the emitted light.

Procedure

Obtain a Boreal spectroscope. While holding the spectroscope with the narrow end at your eye, aim the narrow vertical slit located at the wider end at a light source. Be certain that the scale is to the right of the light source to facilitate viewing the graduations of the scale.

Before you begin the experiment, observe the wavelengths of light emitted by the fluorescent lights in the laboratory. You should observe between two to four distinct spectral lines produced by the fluorescent lights. The spectral lines will be directly under the graduations on the scale. The scale is calibrated in nanometers (10^{-9} meter). For example, you may observe a distinct violet spectral line at 4.3 which should be interpreted as 430 nm. By carefully orienting your spectroscope, you should observe the following spectral lines from the fluorescent lights: violet line at approximately 430 nm, blue line at 475 nm, green line at 560 nm, and possibly a yellow line at 590 nm. The spectroscopes are not accurately calibrated, so you may observe slightly different values than those mentioned here. Also observe the wavelengths of light emitted by sunlight from the nearest window.

This experiment is done in groups of four. There will be 3 stations each containing a different gas and emission wavelengths to observe. Each student in your group should follow the instructions below for making wavelength observations at each station.

Instructions for using emission tubes at each station:

1. Record the name of the gas at the station.

2. Record the color of the emission tube, as viewed from the naked eye.

3. _Do not touch_ the glass emission tubes. _Do not remove_ emission tubes from the power supplies.

4. Turn on the power supply using the red switch on the lower right side of the emission case.

5. View the emission spectrum using your spectroscope. For optimal viewing, stand approximately 1 foot away from the tube.

6. Do not let the emission tube run for more than 1 minute. You may resume viewing by turning the tube on after the unit has been off for 1 minute.

7. Turn off the power using the red switch when are finished viewing the spectrum.

8. Each member of the group should observe the spectrum produced by each emission tube.

9. Record the dominant color of the lit emission tube as observed with the naked eye. Record all spectral lines emitted by the gas at each station as observed with the Boreal spectroscope. All wavelengths should be recorded in units of nanometers (nm).

Data Analysis

When you return to your desk you will choose <u>one</u> of the brightest spectral lines produced by each gas. The wavelengths measured with the spectrometer have units of nanometers. These must be converted to meters in order to calculate frequency, the energy per photon (J/photon) and the energy per mole of photons (kJ/mol). Show your calculations on your data sheet.

Safety Precautions

The emission tubes operate at very high voltages. Exercise care when making your observations and when turning the power supplies on and off. Keep liquids away from the power supplies.

Atomic Emission Prelab Questions

1. What is the frequency of visible light with a wavelength of 500 nm?

2. For each of the following, describe the relationship as directly or inversely proportional:

   a) wavelength vs. frequency

   b) wavelength vs. energy

   c) frequency vs. energy

3. When sodium is excited in a flame, two ultraviolet spectral lines at $\lambda = 372.1$ nm and $\lambda = 376.4$ nm respectively are emitted. Which wavelength represents photons with

   a) higher energy?

   b) longer wavelengths?

   c) higher frequencies?

Atomic Emission Data Sheet

Name _____ **Section** _____

Name of Lab Partner(s) _____

GAS	_____	_____	_____
Color of Emission Tube			
Spectral lines observed λ (nm)			
Frequency of one chosen spectral line υ (sec⁻¹)			
E of the chosen spectral line (J/photon)			
E of chosen spectral line (kJ/mole)			

Sample Calculations

Atomic Emission Review Questions

Name _____ **Section** _____

1. How do your observations of line spectra using the Boreal spectroscope support the concept of quantized energy? (Hint: Compare your observations of the lines observed from the sunlight in the lab window versus your observations of a gas emission tube.)

2. Two emissions of light were observed. One emission (A) was at 410 nm and the other (B) was at 690 nm. Which of these was of greater energy?

3. The first excited state of Ca is reached by absorption of 422.7 nm light. What is the energy difference between the ground state and this excited state?

Atomic Emission Supplemental Questions

Name _____ **Section** _____

Calorimetry

Introduction

Specific heat capacity (usually just called **specific heat**) is an important physical property of a substance that determines how much the temperature of that substance will increase for a given quantity of heat added. The units of specific heat are J/g·°C. *Specific heat (C_s) is defined as the amount of energy required to raise the temperature of 1 gram of a substance 1 degree Celsius under constant pressure.* Water, for example, has an unusually high specific heat (4.184 J/g·°C) and is able to resist temperature changes when heat is added to it. Large bodies of water can absorb large amounts of heat while experiencing only a small change in temperature. This is how the earth's oceans help regulate the temperature of the planet. Conversely, mercury has a relatively low specific heat (0.14 J/g·°C) and is therefore useful as a fluid in thermometers because it responds rapidly to temperature changes.

When heat (q) flows into or out of a substance, the temperature of the substance usually changes.[1] Temperature change can be used to monitor the flow of heat energy. In order to determine the exact amount of heat that flows, we need to know the temperature change the object undergoes (ΔT), the mass of the object in grams (g), and the specific heat of the object (C_s). The mathematical relationship between heat, specific heat, mass, and temperature change is expressed by Equation 1 below.

$$q = C_s \times g \times \Delta T \qquad \text{[Eqn. 1]}$$

The measurement of heat transfer is called **calorimetry.** Heat transfer is usually measured using an insulated container, called a **calorimeter,** which prevents the exchange of heat with the surroundings. The calorimeter you will use in this experiment is called a **coffee cup calorimeter** because it consists of two nested Styrofoam coffee cups. Though far from ideal, this system does a remarkably good job of minimizing the exchange of heat between the calorimeter and the surroundings. A thermometer is used to determine the change in temperature of objects inside the coffee cup calorimeter.

In this experiment, you will determine the specific heat of an unknown metal by measuring the heat transferred when hot pieces of the metal are placed in cold water. Heat will flow from the metal to the water until both are at the same temperature. At this point, the heat lost by the metal (an **exothermic** process, q_{metal} is negative) is equal to the heat gained by the water (an **endothermic** process, q_{water} is positive).

$$+ q_{water} = - q_{metal} \qquad \text{[Eqn. 2]}$$

---

[1] It is possible to add heat to a substance and not have the temperature increase. This is what happens during a phase change. For example, when heat is added to a boiling liquid the temperature of the liquid does not increase, it simply boils more vigorously.

Substituting terms from Equation 1 into Equation 2 leads to the relationship shown in Equation 3.

$$+ C_{s(water)} \times g_{water} \times \Delta T_{water} = - C_{s(metal)} \times g_{metal} \times \Delta T_{metal}$$ [Eqn. 3]

The only unknown term in Equation 3 is the specific heat of the metal. The specific heat of water is a well-known quantity (4.184 J/g·°C) and all of the other terms in Equation 3 represent data collected as part of the experiment.

In the early part of the 19th century, two French scientists published a paper describing the inverse relationship between the specific heats of solid elements and the atomic weights of the elements. The relationship, known as the **Law of Dulong and Petit,** is a simple way to estimate the atomic weight of a solid element if the specific heat for the element is known, or to estimate the specific heat of the element if the atomic weight is known. The data collected in today's experiment will allow you to examine the correlation between inverse of atomic weight and specific heat, just as Dulong and Petit did in 1819.

The Law of Dulong and Petit is a linear relationship where the dependent variable (*y*) is specific heat and the independent variable (*x*) is 1/atomic weight. The slope of the line (*m*) is the proportionality constant in the equation (*y* = *mx* + *b*). The equation for the Law of Dulong and Petit is shown below.

$$C_{s(metal)} = slope \times \frac{1}{atomic\ weight} + y\text{-intercept}$$ [Eqn. 4]

Once you have determined the exact nature of this relationship, you will use it to predict the specific heat of nickel. You will then compare your predicted value for the specific heat of nickel with the value you determine experimentally.

Procedure

You will be assigned a metal sample that will be either aluminum, copper, or lead for which you are to determine the specific heat. Obtain a test tube containing this metal and weigh the stoppered test tube with the metal in it. Pour the metal onto a paper towel (don't loose any!) and weigh the empty test tube with its stopper. Place the metal back in the test tube. Place the test tube, *without the stopper*, into a beaker of boiling water. Make sure the water level is slightly higher than the level of the metal in the test tube. Do not let any water splash over into the test tube. The metal must stay completely dry during this heating process. Heat the metal for 10 minutes to ensure the initial temperature of metal is *approximately* the temperature of boiling water, 100°C.

While the metal is being heated, weigh two nested Styrofoam coffee cups. Add approximately 40 mL of water to the nested cups and weigh them again. Record the mass of the water in the calorimeter cup on your data sheet. Assemble the calorimeter as shown in Figure 1.

Setup your lab computer and MicroLab data acquisition system and prepare it for measuring temperature by opening the experiment file named "Temperature vs. Time" which will provide you with a drop down menu. From this menu you can open the file labeled "Calorimetry and Hess's Law CHML 101". Once the file is opened you will be able to view temperature values in real-time.

Place the temperature probe into the water in the calorimeter and monitor this temperature. Record this value on your data sheet as the $T_{initial}$ of H_2O. Remove the temperature probe from the calorimeter cup and place it in the beaker of boiling water after the metal has been heated for approximately 10 minutes (do not place the temperature probe in the metal in the test tube as this will not give a uniform temperature value). Record this temperature to the nearest 0.1°C on your data sheet as $T_{initial}$ of metal. Remove the temperature probe from the boiling water and wipe it dry. When it has cooled to room temperature, approximately 22–25°C, place the temperature probe in the calorimeter cup.

When you have finished heating for the specified period, turn off the Bunsen burner flame. Click the "Start" prompt on the computer program to begin collecting temperature data per second. Using a test tube holder or pot holder,

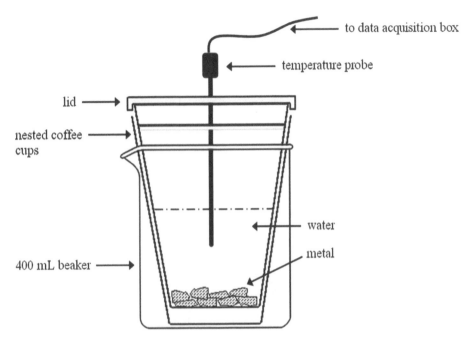

FIGURE 1 — *Coffee Cup Calorimeter*

remove the test tube containing the heated metal from the boiling water and quickly empty the contents into the calorimeter. Be careful that no water adhering to the tube falls into the calorimeter. Cover the calorimeter and stir the contents with the temperature probe being careful not to touch the metal pieces. Refer to Figure 1 as a reference. Continue to collect temperature data for an additional three minutes. When the temperature has reached a high point and begins to return to room temperature, you may click the "Stop" prompt. Record the highest temperature value as the final value of temperature (to the nearest 0.1°C) in the line for T_{final} of H_2O and metal on your data sheet.

This procedure will be repeated with the nickel metal. Obtain a test tube of nickel metal and repeat the procedure outlined above for determining specific heat.

When you have finished both the assigned metal (aluminum, copper or lead) and the nickel specific heat determinations, you must *dry* the metals. This is done by placing the metal in its test tube without the stopper back in the boiling water bath. Once the metal is hot, pour it out on a dry paper towel. Dry the inside of the test tube with a paper towel. Wet and/or corroded metals will not give correct specific heat data.

Make sure that the Bunsen burner, wire gauze and ring clamp are cool before disassembling and returning them to their designated locations. Make certain that each metal test tube is stoppered with the correct label for that specific metal, and that all metals are completely *dry before you return them to their proper location in the container on the front reagent bench*.

Safety Precautions

Remove the stopper on the test tube when heating your metal in boiling water. Flame of Bunsen burner must be *off* when retrieving test tube from heated water bath. Wire gauze, ring clamp, and Bunsen burner must be *cool* before disassembly to avoid serious burns!

Waste Management

Return calorimeter cups, lids and *dried* metals to the front reagent bench.

Analysis of Results

A. To determine the ΔT H_2O, find the difference between the T_{Final} H_2O + metal and $T_{Initial}$ H_2O. To determine the ΔT metal, find the difference between T_{Final} H_2O + metal and $T_{Initial}$ metal (note that this value will be negative). Use these values in Equation 3 to determine the specific heat of your assigned metal and record this value, along with the identity of the metal, in the space provided on the whiteboard in the lab and on your data sheet.

B. Determine the specific heat of nickel using Equation 3. Record this value on your data sheet.

C. When all of the lab groups have reported the values of the specific heat of their assigned metals (aluminum, copper and lead specific heat values), plot these values on Graph 1 supplied on the data sheet which plots *specific heat vs. 1/atomic weight*. Draw a best-fit straight line using a ruler through the clusters of data points.

   Your instructor may ask you to provide your lab data to incorporate into an Excel Worksheet. From this worksheet your instructor can display the graph of *specific heat vs. 1/atomic weight*. Your Instructor can demonstrate the best-fit line and the slope of your line in the Excel program. You may use this worksheet table and graph in your Calorimetry Lab Report after your instructor has emailed this to the class.

D. Use Graph 1 to determine the specific heat value of nickel, knowing that its atomic weight is 58.7 g/mol. Record this value on your data sheet.

E. Determine the value of the slope for your graph and substitute this value for the slope in Equation 4. Use this equation, derived from the Law of Dulong and Petit, to determine the specific heat of nickel and record this value on your data sheet.

F. Compare the experimentally determined value for the specific heat of nickel to the value you obtained by graphical analysis. Record the % difference between these two values on your data sheet using equation 5 below.

$$\% \text{ difference} = \frac{(\text{experimental value} - \text{predicted value})}{\text{experimental value}} \times 100 \qquad \text{[Eqn. 5]}$$

G. Formulate a draft conclusion in which you interpret and compare the results of the various metal specific heat determinations. This should include discussing the principles learned from the experiment and evaluative comments of how the results corroborate with these principles. Discuss all types of data (graphs, tables, etc.) to explain and relate your results to the basic principles learned about calorimetry. How might you explain any unexpected results or sources of error? Discuss these possible sources of error in your draft.

H. Discuss your draft conclusion with your instructor before you leave lab to make sure that you understand the underlying concepts of the experiment, how to present and discuss your results in a formal lab report, and how to generate a good conclusion. Your instructor will sign the draft if satisfactory.

Calorimetry Prelab Questions

Name _____ **Section** _____

1. How much thermal energy (q) is required to heat 15.2 g of a metal (specific heat = 0.397 J/g·°C) from 21.0°C to 40.3°C?

2. 100 g of water (specific heat = 4.184 J/g·°C) and 100 g of a metal sample (specific heat = 0.397 J/g·°C) are heated from 25°C to 75°C. Which substance required more thermal energy to reach 75°C? (Explain your reasoning).

Calorimetry Data Sheet

Name _____ **Section** _____

Name of Lab Partner(s) _____

	Metal Sample	Nickel
1. Metal used / atomic weight	_____	_____
2. Weight of test tube + metal	_____	_____
3. Weight of test tube	_____	_____
4. Weight of metal	_____	_____
5. Weight of calorimeter + H_2O	_____	_____
6. Weight of calorimeter	_____	_____
7. Weight of water	_____	_____
8. $T_{iniital}$ of H_2O	_____	_____
9. $T_{initial}$ of metal	_____	_____
10. T_{final} of H_2O + metal	_____	_____
11. $\Delta T\ H_2O$	_____	_____
12. ΔT metal	_____	_____
13. Specific heat value for metal determined by experiment	_____	_____
14. Specific heat of Ni predicted using Equation 4	_____	_____
15. Percent difference of nickel values	_____	_____

Class Results

Table 1 *Atomic Weight and Specific Heat Information for Certain Metals*

Metal	Atomic Weight (g/mol)	1/Atomic Weight (mol/g)	Specific Heat Values (J/g·°C)
Cu	63.54	0.0157	
Al	26.98	0.0371	
Pb	207.19	0.0048	
Ni	58.69	0.0170	(determined from graph below)

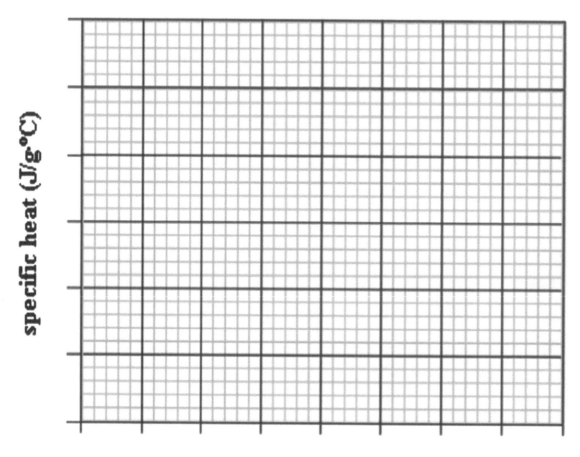

GRAPH 1—*Relationship of Specific Heat vs. Inverse of Atomic Weight*

Conclusion Discussion

Formulate a draft conclusion in which you interpret and compare the results of the various metal specific heat determinations. Write your draft in the space below. When you have completed your draft, have your instructor review and discuss your statement and sign on the designated line below. This page will be turned in with your Calorimetry Lab as part of your grade.

Instructor's Signature: _____

Calorimetry Lab Report

For progress to occur, researchers must convey the significance of their research to the scientific community and to the general public. Conveying theories and data in an understandable, concise, and convincing manner will help their research gain better acceptance and enable future ideas build upon those newly formed. To achieve success in almost every career, not just in science, the ability to communicate effectively and accurately by writing is essential.

As a University Studies course that explores the *Scientific Analysis of the Natural World*, general chemistry 101 laboratory has adopted student learning outcomes designed to help students attain specific skills in inquiry and thoughtful expression. These outcomes are not only an integral part of learning science, but an integral part of a fulfilling college education. To achieve these outcomes, CHML 101 students are required to prepare a written report of the Calorimetry lab. Writing a formal lab report gives students the opportunity to practice writing skills needed to explain important ideas and findings.

The Calorimetry lab report involves writing a concise introduction and purpose of the experiment, and documentation and explanation of data gathered. Students will write a justification of their results, and discuss possible sources of error in the conclusion section of the report. Students will also use their new knowledge of the subject to answer summary questions in essay form, using their own results to aid in the discussion. The report should not exceed 5 pages, double spaced, and will be due at the beginning of the next week of lab class. Reports not turned in during the regularly scheduled lab class will *not* be accepted (no reports accepted late). The Calorimetry lab report is worth up to 45 points and *cannot be dropped* even if it represents the lowest lab score. The report replaces the lab grade for the Calorimetry experiment; therefore students will not turn in their pre-lab questions, data sheets, and review questions until the report is complete and handed in the following week. *There will be no exceptions to this grading policy.* The report will be evaluated on the following criteria:

1. **Scientific explanation.** The report should be clearly written, scientifically sound, and show a thorough understanding of the objectives and results of the experiment.

2. **Adherence to the lab report writing convention outlined below.** Demonstrate the proper presentation of formal lab reports (modified as outlined below) with heavy emphasis on descriptions and explanations of observations, calculations, data, tables and graphs.

3. **Demonstration of professional writing.** The report should demonstrate good grammar, spelling, word usage, and comprehension. Points will be deducted for incomplete sentences (except for graph and table labels). The report should be written so a non-science student could understand the basic principles and explanations of data and conclusions.

The following format should be followed in writing the Calorimetry report. Note that it is a modification of a typical formal lab report. The paperback book, *A Short Guide to Writing about Chemistry* by Davis, H., Tyson, J. and Pechenik, J. (Pearson, Inc.; 2010), may be used as an instructional guide for preparing laboratory reports.

A.   **Pre-Lab questions**

B.   **Title and Author.** The title should be straightforward and reflect the content of the report. Include your name and pertinent information (section number, TA name, date) on the report front cover.

C.   **Introduction.** Provide an explanation of the definition of calorimetry, why it is important, and what it is used for. Briefly discuss the main objectives and goal of the experiment. Briefly describe the general experimental design or method. (The Introduction section should be written in the *present* tense.)

D.   **Materials and Procedure.** Students may cite the Lab Manual for this section, unless the procedures were changed, or different materials were used than those described in the lab manual. Students must describe the procedures and materials actually used, if these deviated from the manual. (The Procedure should be written in the *past* tense.)

E.   **Results and Discussion.** Presentation of findings should clearly describe and explain the data and results. Tables and graphs must be used. (The Results section should be written in the *past* tense.)

F.   **Conclusion and Discussion.** In this section the results must be interpreted and compared in relation to the specific purpose of the experiment. This should include discussing the principles learned from the experiment and evaluative comments of how the results corroborate with these principles. Relevant work in the literature can be added. The Conclusion section should end with a short summary regarding the significance of the work. Use all types of data (graphs, tables, etc.) to explain your conclusion and relate your results to existing theory and knowledge. (You may include or attach your data sheet with tables and graphs and/or the excel worksheet from your instructor to your report.) How might you explain any unexpected results? Attach Review Questions to your report if requested by your instructor.

G.   **Instructor's Signature.** Attach your draft conclusion with your instructor's signature in your report.

H.   **Questions.** Form conclusions about the following 2 essay questions by *incorporating the results of the experiment* to justify the answers.

   1. Explain what might happen to your results in the Calorimetry experiment if the calorimeter cup did not have a lid.

   2. How would your results have varied if you put in half the volume of water in the calorimeter cup?

I.   **References.** Students must cite at least 2 outside sources in addition to the General Chemistry lecture book and Lab Manual. Websites such as Wikipedia are acceptable as references **only if the original source of the information is cited**. The listing should be alphabetized by the last names of the authors. Some examples are as follows:

   1. Citing References within Text of Report

       When citing references in the paper refer to information by the author's name and date the information was published. For examples:

       a) "Harris in 2009 determined the caffeine in regular coffee to be 106–164 mg per 5 oz serving size."

       b) "Caffeine in regular coffee has been determined to be 106–164 mg per 5 oz serving size (Harris, 2009)."

   2. Books

       Harris, D.C. 2009. *Exploring Chemical Analysis, 4th edition.* New York: W.H. Freeman and Co.

   3. Chapters in Books

       Harris, D.C. 2009. Chemical Measurements. In *Exploring Chemical Analysis, 4th edition.* Pp.18–33. New York: W.H. Freeman and Co.

   4. Journal Publications

       Kieber, R.J., Whitehead, R.F. and Skrabal, S.A. "Photochemical Production of Dissolved Organic Carbon from Resuspended Sediments" *Limnol. Oceanogr.* 2006, *51*, 2187–2195.

5. Websites

   Kugler, W. X-Ray Diffraction Analysis in the Forensic Science:

   The Last Resort in Many Criminal Cases, http://www.icdd.com/resources/axa/vol46/V46_01.pdf (accessed January 19, 2012).

6. Laboratory Manuals or Handouts

   Nelson, J., Kemp, K., Stoltzfus, M., Gravimetric Determination of Phosphorus in Plant Food, *Laboratory Experiments*, Pearson Education, Inc., 2012; pp 103–113.

Calorimetry Review Questions

Name _____ **Section** _____

1. If the specific heat of a metal is 0.850 J/g·°C, what is its atomic weight? (Hint: use the Law of Dulong and Petit.)

2. One way to cool a cup of coffee would be to plunge an ice-cold piece of aluminum into it. Suppose a 20.00 g piece of aluminum is stored in the refrigerator at 0.0°C and then dropped into a cup of coffee. The coffee's temperature drops from 90.0°C to 75°C. How many kJ of thermal energy did the piece of aluminum absorb? (Specific heat$_{Al}$ = 0.902 J/g·°C.)

3. Using information from question #2 above, calculate the mass (g) of coffee in the cup. (Specific heat$_{coffee}$ = specific heat$_{water}$ = 4.18 J/g·°C.)

4. Between 0°C and 30°C, mercury has a specific heat of 0.138 J/g·°C. If 200 J of heat are removed from 100 g of Hg at 20°C, what will the final temperature of the Hg be? (Hint: this is an exothermic process because the Hg is losing heat.)

Calorimetry Supplemental Questions

Name _____ **Section** _____

Chromatography I: Paper

Introduction

In 1906, a Russian botanist, Michael Tswett, devised a new and ingenious method for the separation of plant pigments. By pouring an extract of leaf material through a vertical glass tube packed with powdered chalk, he observed that the extract separated into several distinct, colored zones. He named his process "chromatography". He was further able to extract the materials from each zone by pressing the chalk from the tube, cutting between zones, and extracting each with alcohol. By this technique, he had separated a complex mixture into its components.

The technique of chromatography has found wide application in chemistry. It is not necessary to work with colored substances. For example, colorless amino acids are frequently separated by chromatography and a dye or special lights are required to locate the separated components on the chromatogram. The original column chromatography has led to such newer methods as paper chromatography, thin-layer chromatography, vapor-phase chromatography, high performance liquid chromatography, and ion-exchange chromatography. It is important to realize these methods share a common characteristic with the original column method: *they all separate mixtures into individual components based on the physical property of relative solubility of components between a stationary phase and a mobile phase.*

In this experiment you will apply the theories of chromatographic separation using **paper chromatography**. Paper chromatography is an example of a liquid-liquid *physical partitioning method*. The **solutes** in a liquid mixture are separated, or **resolved**, by being physically moved between two phases because of solubility differences based on their differing chemical properties. The method is not a chemical separation technique, as it does not alter the composition of the substances.

Cellulose filter paper is capable of holding in its pores a layer of water, which it absorbs from moisture in the air. This water layer acts as the **stationary phase** because it is trapped in the filter paper. When one end of the filter paper is dipped into a liquid solvent, the solvent is drawn up the paper by capillary action. This liquid solvent is termed the **mobile phase**, as it is moving rapidly, flowing over the paper-water surface. As the solvent ascends, the solute compounds originally spotted on the filter paper will move with it.

Generally, the mobile phase will have physical properties differing slightly from the water in the stationary phase. The type and composition of the solvent used is varied in order to produce the best resolution of the components in a mixture. If a substance in the mixture has properties similar to the water layer, then the substance will prefer to remain dissolved in the water layer attached to the paper. On the other hand, if the substance has properties similar to the solvent of the mobile phase, it will dissolve in the mobile phase and be carried with it up the paper.

Solutes with more affinity for water in the stationary phase will not travel very far with the mobile solvent and will move more slowly up the paper. Those solutes that are more soluble in the mobile phase will be carried up the paper farther, possibly to the front edge of the rising solvent called the **solvent front**. Eventually, the substances in the mixture will be separated into diffuse spots away from the position where they were originally spotted, called the **origin** or the **baseline**.

Each component of a mixture is held back by the stationary phase to some extent. In addition, each component has a varying solubility in the mobile phase causing each to move forward, up the paper, with the solvent to a different extent. Therefore, partitioning of compounds in a mixture occurs due to the individual *solubilities* in the mobile phase versus the stationary phase. As substances are generally slightly soluble in both phases, there is a continuous back-and-forth exchange of solutes between the water and the solvent. This combination of being held back and moving forward results in the physical separation of the components in the mixture.

The resulting location of a component on the chromatogram is characteristic of that component when measured under specific conditions. A term called the **retardation factor, R_F**, is used to describe the location of a component on the chromatogram. The retardation factor is calculated by dividing the distance traveled by a particular solute by the distance traveled by the solvent.

$$R_F = \frac{\text{distance traveled by solute}}{\text{distance traveled by solvent}}$$

The distance traveled by the solute is measured from the **origin** (baseline) to the center of the component spot on the chromatogram (distance A or B in Figure 1). The distance traveled by the solvent is measured from the origin to the **solvent front** (distance C in Figure 1). There are no units on the R_F value as it is the ratio of two numbers with the same units. Note that the R_F value must be ≤ 1.

The R_F value for a solute is a physical property that can be used to identify a substance when it is separated from a mixture of solutes on a chromatogram. The exact conditions of the chromatographic experiment must be stated when reporting R_F values as many factors can affect the relative solubility of the solute in the stationary and mobile phases. Specific factors include solvent system, type of paper, and temperature.

The object of today's experiment is to become familiar with separations by paper chromatography and relate these to the intermolecular forces of solutes and solvent. You will use this technique to determine R_F values for several organic acid/base indicators and to determine the composition of a mixture of these indicators. The colors of these dyes may alter depending upon whether they are in a slightly acidic or basic environment. To increase the accuracy of your identifications, you will expose your chromatogram to ammonia vapors. This will cause the dyes to change color, making them easier to locate and identify. These are the colors you are to record on your data sheet.

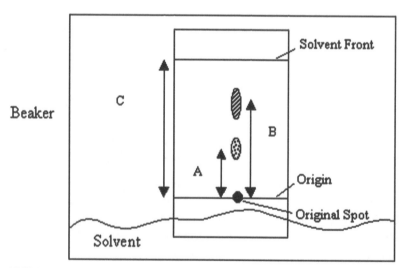

FIGURE 1 — *Sample Paper Chromatogram*

Procedure

Obtain 25 mL of chromatography solvent (a mixture of butanol, ammonia, and water) and pour it into a clean, dry 600 mL beaker. Immediately cover the beaker with a piece of plastic film and fasten it with a rubber band. Allow the beaker to stand at least 10 minutes in order to saturate the air in the beaker with solvent vapors. In the meantime, prepare your chromatography paper.

Obtain a sheet of 10 cm × 20 cm chromatography paper and using a pencil, draw the origin line about 1 cm from the longer edge of the paper. Handle the paper as little as possible and lay it on the bench. Using the capillary tube supplied with each indicator (do **not** remove the indicator bottle tops) apply a small spot (about 2 mm in diameter) of each indicator on the pencil line. Make sure the spots are at least 2 cm apart. Write the name of each indicator in pencil below its spot.

Be careful to return the capillary tubes to the proper bottles to avoid contaminating the indicators. Now, select one of the unknown mixtures and spot it on the pencil line. Be sure to record the number of the unknown on your data sheet. After you have made all the spots and allowed them to dry, carefully re-spot each one to increase the concentrations.

When the spots have dried, form the paper into a cylinder with the spots facing out. Do not overlap the edges. Staple the edges together with two staples making sure that the edges are even and not touching. Remove the plastic wrap from the 600 mL beaker containing the solvent and quickly place the paper cylinder into the beaker with the origin edge dipping into the solvent. The spots *must* be above the solvent level, otherwise the indicators will be leached out into the solvent. Replace the plastic wrap and fasten it with the rubber band. Allow the chromatogram to develop, undisturbed, until the solvent front is about 2 cm from the upper edge of the cylinder. Remove the cylinder and immediately mark the solvent front with a pencil. This must be done while the front is still wet. Allow the chromatogram to dry before proceeding.

When the chromatogram has dried, take it to the main fume hood. Moisten the chromatogram (don't soak it) with water from the mister and hold it over the ammonia tank (a large beaker with concentrated ammonia in the bottom) for 10 seconds. Remove the chromatogram and unroll it. Immediately circle with a pencil the major individual spots of color that have separated. Some indicators may contain impurities and/or decomposition products, which will appear as minor spots of color.

Using the estimated center of the circled spot, compute R_F values for each major individual color from the known dyes on the chromatogram. Compare the R_F values to determine the order of polarity for the known dyes, designating 1 as the most polar, and 5 as the least polar compound.

Record the number of your unknown mixture on your data sheet. Identify the components of your unknown mixture by comparing the R_F values and colors of the unknown resolved solutes with those of the known dyes. Compare the R_F values of your unknown mixture to assign an order of polarity, where 1 is the most polar solute, and 3 is the least polar solute within your unknown solution. These designations should correspond with the order of polarity assigned to your standards. When you have finished, staple your chromatogram to your data sheet or your lab partner's data sheet.

Safety Precautions

This experiment involves the use of butanol and concentrated aqueous ammonia. Butanol is toxic if ingested and extremely irritating to the eyes. Ammonia vapors are very irritating to the lungs, mucous membranes, and eyes. Exercise caution when handling these materials. Wash thoroughly after handling and keep butanol containers covered. Both the butanol/ammonia chromatography solvent and the ammonia developing tank are located in the fume hood.

Waste Management

After you complete the experiment, ask your Instructor to check the quality of your Butanol/Ammonia chromatography solvent. If your Instructor determines that your Butanol/Ammonia chromatography solvent is *not* contaminated, please pour the solvent into the bottle labeled "**RECYCLED Butanol Solvent**" (Mobile Phase solvent for paper chromatography) in the fume hood once your experiment is complete. DO NOT add water to this bottle if you are trying to wash out your beaker!!

If your Instructor determines that your chromatography solvent *is contaminated*, please pour the remaining solvent into the container labeled "Organics waste" container located in the fume hood.

Chromatography I: Paper
Prelab Questions

Name _____ **Section** _____

1. Observe the following picture of several known solutes spotted on the stationary phase, phenol red, bromcresol purple and bromothymol blue respectively. Which solute spot would most likely cause problems with the chromatography process and why?

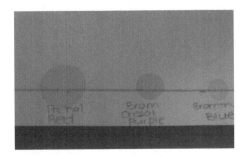

2. Answer the following questions using Figure 1 from the Introduction section of this Laboratory:

   a) Measure to the correct number of significant figures the distances A, B and C and calculate the R_F values for the two components that separated from the original mixture. You may measure A, B and C in any units that are convenient for you.

   b) R_F for   ⬭   _____

   c) R_F for   ⬭   _____

   d) Considering the two solutes in Figure 1, which substance is most soluble in the mobile phase?

   e) Which substance is least soluble in the mobile phase?

   f) Which substance in Figure 1 is more polar, given that the stationary phase is water held within the pores of the cellulose paper, and the mobile phase is butanol (C_4H_9OH)?

   g) Following the same logic as in question f above, which substance in Figure 1 is the more non-polar compound?

Chromatography I: Paper Data Sheet

Name _____ **Section** _____

Name of Lab Partner(s) _____

Analysis of Known Indicators

Indicator	Color in Ammonia	R$_F$ Value	Order of Polarity
Alizarin Yellow	_____	_____	_____
Bromcresol Purple	_____	_____	_____
Bromthymol Blue	_____	_____	_____
Phenol Red	_____	_____	_____
Phenolphthalein	_____	_____	_____

Analysis of Unknown Mixture

Unknown Number _____

Colors of Spots in Unknown Mixture (After development in Ammonia Tank)	R$_F$ Values	Order of Polarity
1. _____	_____	_____
2. _____	_____	_____
3. _____	_____	_____

Unknown mixture # _____ contains the following known dyes:

1. _____   2. _____   3. _____

Chromatography I: Paper
Review Questions

Name _____ Section _____

1. How would you expect the R_F value to change for a given set of solutes as their solubility in the mobile phase increases? Explain.

2. Organic dyes A, B and C are separated from their mixture by paper chromatography. Their R_F values respectively are: 0.92, 0.35 and 0.68. Which dye is the most soluble in the mobile phase? Which is the most soluble in the stationary phase?

3. Two mobile phases were evaluated by a student to study the behavior of five dyes. The following data from each chromatogram were collected:

	Mobile Phase 1: *Acetone/water*		**Mobile Phase 2:** *Ethanol/water*	
	Distance Traveled (cm)	R_F	Distance Traveled (cm)	R_F
Solvent front	4.7	—	4.9	—
Green dye	2.9		4.8	
Red dye	1.3		3.7	
Yellow dye	4.0		0.8	
Blue dye	2.9		2.6	
Purple dye	3.1		1.4	

a) Calculate the R_F values for each food dye for both chromatographic systems. Record these values in the table above.

b) Explain your reasons for choosing which solvent would be most effective in separating the green from the blue dye.

c) Which mobile phase system would be most effective for separating all 5 dyes? Explain.

Chromatography I: Paper
Supplemental Questions

Name _____ **Section** _____

Chromatography II: HPLC

Introduction

There are many different types of chromatography, all of which are based on the same fundamental principles. *Compounds of interest (analytes) are separated from one another by their differing affinities for a stationary phase and a mobile phase.* Different types of chromatography arise from the types of stationary and mobile phases used and whether the stationary phase is in a film on an inert sheet (planar chromatography) or if it is in a tube (column chromatography). The paper chromatography experiment performed earlier was an example of liquid-liquid planar chromatography because both the stationary and mobile phases were liquids. In this experiment, you will use a type of column chromatography called **High Performance Liquid Chromatography** (HPLC) to separate and quantify caffeine from other organic compounds in soft drinks.

HPLC is performed by injection of a small quantity of sample into a moving stream of liquid, the **mobile phase**. The mobile phase in this experiment is 49% methanol (CH_3OH) and 51% water. The desired polarity for best separation of analytes in a mixture determines the composition of the mobile phase. The mobile phase is pumped at a uniform rate at high pressure through a column packed with particles, which serves as the **stationary phase**. The stationary phase in the HPLC instrument you will be using is an alkane that has been bonded to the surface of silicon dioxide spheres. Each sphere is 5 μm in diameter and is covered by alkane molecules containing 18 carbons, thus giving the column its name, "C-18". The alkane-coated silicon dioxide spheres are called **packing** and are very nonpolar.

Separation of a mixture into its individual components depends on the differing degrees of affinity of each solute for the C-18 column. The amount of time an individual solute stays attracted to the column is dependent on its partitioning between the liquid phase pumped steadily through the column and the stationary phase in the column. Compounds with similar intermolecular forces have a greater affinity for each other and want to remain together. Thus, analytes will elute more rapidly if they are more soluble in the mobile phase and will elute more slowly if they are more soluble in the stationary phase.

The block diagram in Figure 1 shows the main components of the HPLC instrument. The very small diameter of the packing particles filling the column, and the need to move the analytes through the column and detector quickly mean the capillary action used in the paper chromatography experiment is no longer a satisfactory method to move the mobile phase over the stationary phase. Therefore, a high-pressure **pump** is needed to force the mobile phase through the column, typically at pressures of 500–5000 pounds per square inch (psi). **Degassing** the mobile phase is critical because gas bubbles cause problems in the pump, column, and detector. Because forcing the mobile phase through the column occurs at high pressure, a special **injector** is needed to introduce small quantities of sample (10–500 μL) into the column. The **filter** and **precolumn** are used to prolong the life of the column. The filter traps small particles of contaminants before they can reach the column. The precolumn, which contains a small amount of the same type of packing found in the column, traps compounds that interact

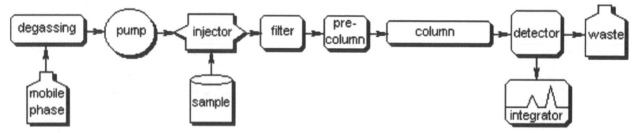

FIGURE 1 — *Block Diagram of HPLC*

so strongly with the column that they cannot be removed by the mobile phase. If these small particles and compounds were allowed to build up at the front of the column, it would become clogged and unusable much sooner which is significant because columns can cost from several hundred dollars to over a thousand dollars.

Analytes, which have been separated from the mixture, are allowed to exit the column and flow through a detector. The time it takes an individual compound of a sample mixture to travel from the injector to the detector is called the **retention time**, which is used in the same manner as the R_F (retardation factor) in the paper chromatography experiment. To identify individual solutes in a sample mixture, it is necessary to compare retention times of unknown analytes with retention times of standards that have already been injected into the HPLC. As with paper chromatography, the retention time of a particular analyte is constant for a fixed set of chromatographic conditions such as flow rate, temperature, type of column, and mobile phase composition.

In addition to qualitative information, a **detector** provides quantitative information regarding the concentration of the analyte. The UV-Visible detector used in this experiment is very similar to the Spectronic 20 instrument you may have used in previous experiments in which you learned Beer's Law and the correlation between concentration of a substance and its absorbance at certain wavelengths ($A \propto C$). The wavelength used in this experiment is 274 nm, which is where the maximum absorbance of caffeine occurs.

In the course of diffusion through the HPLC column, all the analyte does not reach the detector at once. You observed this same phenomenon in the paper chromatography experiment when the small dot of sample applied at the bottom of the paper spread out as it moved up the paper. The leading and trailing edges were lighter in color (lower concentration) than the middle portion, which was darker (higher concentration). As the analyte moves through the detector, the absorbance changes as the less concentrated leading edge gives way to the more concentrated middle portions, which in turn give way to the less concentrated trailing edge. This causes a peak shaped response for the analyte composed of many different absorbance values. These values are sent to the **integrator** which sums up (integrates) all the absorbance values for each peak. Because absorbance is directly proportional to concentration, the integrated signal is directly proportional to the concentration of caffeine. By analyzing a set of standards of known concentrations, a calibration curve can be constructed of peak area (detector response) versus concentration of analyte.

The structure of caffeine (mol.wt. = 194.19) is shown in Figure 2 below. It is a relatively nonpolar molecule due to the methyl groups (CH_3) branching out from the structure. Using the "like dissolves like" principle, caffeine is more soluble in a less polar solvent and relatively less soluble in a more polar solvent like water.

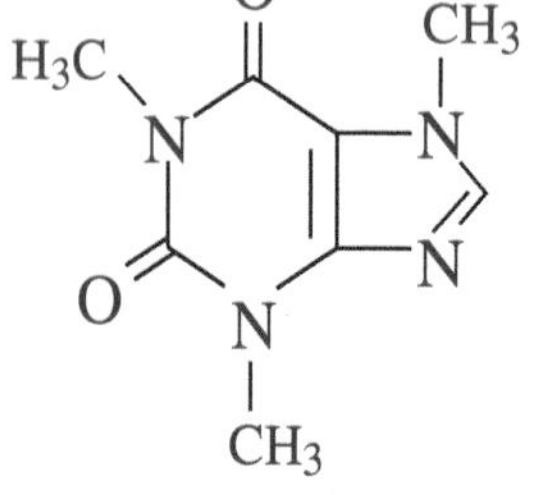

FIGURE 2 — *Structure of Caffeine (mw = 194.19)*

Procedure

This experiment is to be performed in pairs. In Part I you will prepare four standard caffeine solutions. A **standard** is an established quantity and quality of a substance. In this experiment standards will be made consisting of pure caffeine in DI water. First, a **stock solution** of pure caffeine will be prepared from which you will make your standards. The stock solution is generally made at a higher concentration that requires a significant mass of caffeine crystals to ensure better accuracy and precision. From this stock solution you will prepare standards that are in an optimal concentration range for analysis by the HPLC instrument.

In Part II you will prepare a sample from a soft drink of your choice that will be run on the HPLC using an auto sampler where samples are withdrawn, injected, and analyzed automatically and the results printed for you. While the samples are running, your instructor will help you identify the different parts of the chromatograph in the schematic in Figure 1 and develop a standard calibration curve using chromatograms of standard caffeine solutions.

Part I: Preparation of Standard Caffeine Solutions

A. PREPARING A STOCK 5.00×10^{-3} M CAFFEINE STANDARD

Obtain a 250 mL **volumetric flask** and clean it well with DI water. A volumetric flask is a pear-shaped glassware that has a flat bottom and is made to have a narrow neck to be more precise. The volumetric flask is calibrated to contain a specified volume of liquid and is marked near the center of the neck to indicate that specific volume. Volumetric glassware, including flasks and pipets, are calibrated to approximately ± 0.01 to ± 0.05 mL. When preparing a solution, the meniscus of the liquid should be on the calibration mark.

Add a small amount of DI water to the volumetric flask. Make a crease in your weighing paper by folding it in half. Tare the weighing paper on the electronic balance and weigh out 0.240 g of pure standard caffeine. Carefully roll the weighing paper into a cylindrical shape and insert the tip of the weighing paper into the neck of the volumetric flask. Deposit the caffeine crystals into the volumetric flask, and using your squeeze bottle filled with DI water remove any residual crystals from the weighing paper, flushing them into the flask. Swirl the volumetric flask to help dissolve the crystals in the small amount of DI water already in the flask. Fill the flask with DI water until the water level is close to the calibration mark. Make sure you swirl the contents after small additions of DI water. Using your squeeze bottle, add the remaining DI water to the volumetric flask so that the meniscus lies on the calibration mark. Place and hold the cap on the volumetric flask while your swirl the contents to dissolve the caffeine. Hold the cap on the flask tightly and shake the flask until all of the caffeine is dissolved. This is your stock solution of 5.00×10^{-3} M caffeine.

B. PREPARING CAFFEINE STANDARDS FROM DILUTIONS OF STOCK STANDARD

You will prepare several caffeine standards (solutions of known concentration) by diluting the stock caffeine solution (5.00×10^{-3} M) with water. The standards will be prepared in 50 mL volumetric flasks. The following formula can be used to determine what volume is necessary to pipet from the stock solution to make a particular standard.

$$M_1 V_1 = M_2 V_2 \qquad \text{[Eqn. 1]}$$

Where M_1 is the molarity of caffeine in the newly prepared (diluted) solution, M_2 is the molarity of caffeine in the stock solution (5.00×10^{-3} M), V_1 is the volume of the diluted caffeine solution (this is always 50.00 mL in this experiment). Solving the equation for V_2 gives the volume of stock solution necessary to remove from the stock solution to prepare the new solution of concentration M_1. For example, to prepare a new standard of 5.00×10^{-4} M caffeine, 5.00 mL of stock solution is necessary to pipet into a 50 mL volumetric flask.

Using Equation 1, calculate the volume of 5.00×10^{-3} M caffeine stock solution (M_2) necessary to pipet into a 50 mL volumetric flask to make the following caffeine standards: 2.50×10^{-4} M, 1.00×10^{-4} M, and 0.500×10^{-4} M. Enter these values for V_2 in **Table 1** in the **Prelab Questions** and have your instructor check these values before preparing your standards.

Pour the stock solution, M_2, into a small, clean and dry beaker so that the solution is easier to withdraw by pipet. Rinse the pipet four times with small volumes (2–3 mL) of the stock caffeine sample before dispensing the appropriate volume (V_2) into the volumetric flask. Fill each 50 mL volumetric flask up to the calibration mark with deionized water using your wash bottle (plastic squeeze bottle). The meniscus should be right on the calibration mark. These are your standard caffeine solutions (M_1).

Standards analogous to the ones you prepare have been analyzed by HPLC and their chromatograms are shown in Figures 3, 4, 5, and 6. **A chromatogram** is a graph of signal (peak area in this experiment) plotted on the y-axis versus time on the x-axis. Record the caffeine retention times and peak areas from the chromatograms in Figures 3, 4, 5, and 6 on your data sheet.

Caffeine Standard 1: 0.500×10^{-4} M

Injection Date   :
Sample Name   :
Inj Volume   : 10 µl

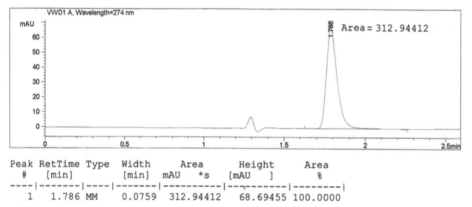

Peak #	RetTime [min]	Type	Width [min]	Area mAU *s	Height [mAU]	Area %
1	1.786	MM	0.0759	312.94412	68.69455	100.0000

FIGURE 3 — *Caffeine Standard 1: 0.500 × 10⁻⁴ M*

Caffeine Standard 2: 1.00×10^{-4} M

Injection Date   :
Sample Name   :
Inj Volume   : 10 µl

Peak #	RetTime [min]	Type	Width [min]	Area mAU *s	Height [mAU]	Area %
1	1.774	MM	0.0754	616.07892	136.21225	88.7791

FIGURE 4 — *Caffeine Standard 2: 1.00 × 10⁻⁴ M*

Caffeine Standard 3: 2.50 x 10⁻⁴ M

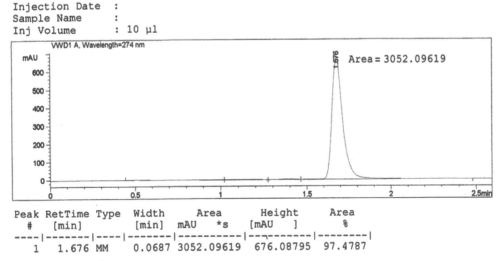

Injection Date :
Sample Name   :
Inj Volume   : 10 µl

```
VWD1 A, Wavelength=274 nm
mAU                                                    Area = 1603.80884
300
250
200
150
100
 50
  0
   0        0.5         1          1.5          2            min
```

```
Peak RetTime Type  Width      Area        Height      Area
  #   [min]        [min]  mAU    *s    [mAU    ]       %
----|-------|----|-------|----------|--------------|--------|
  1   1.723  MM   0.0700 1603.80884  346.53311     95.5767
```

FIGURE 5 — *Caffeine Standard 3: 2.50 × 10⁻⁴ M*

Caffeine Standard 4: 5.00 x 10⁻⁴ M

Injection Date :
Sample Name   :
Inj Volume   : 10 µl

```
VWD1 A, Wavelength=274 nm
mAU                                                    Area = 3052.09619
600
500
400
300
200
100
  0
   0        0.5         1          1.5          2         2.5min
```

```
Peak RetTime Type  Width      Area        Height      Area
  #   [min]        [min]  mAU    *s    [mAU    ]       %
----|-------|----|-------|----------|--------------|--------|
  1   1.676  MM   0.0687 3052.09619  676.08795     97.4787
```

FIGURE 6 — *Caffeine Standard 4: 5.00 × 10⁻⁴ M*

Part II: Preparation of Soft Drink for Analysis

Rinse a 10 mL volumetric pipet with several portions of deionized water and transfer 10 mL of water into a clean dry 50 mL beaker using this pipet. Choose a decarbonated soft drink of your choice from those on the counter and record the type on your data sheet. Pour approximately 15 mL of sample in a clean dry 100 mL beaker. Rinse a volumetric pipet four times with small volumes (2–3 mL) of soft drink sample and then pipet 10 mL of sample into the 10 mL of DI water. Swirl the contents of the beaker to ensure it is well mixed. Once your sample dilution is complete, your instructor will give you a chromatogram of your soda sample. The identity of the caffeine peak in your soda sample chromatogram is based on closest similarity to the retention time of the caffeine peak in chromatograms of pure standards.

Determination of Caffeine Content in Soft Drink

Record the values of retention time and corresponding peak area for each standard on your data sheet in Table 2: Information for Standards. Using this information, prepare a calibration curve on graph paper by plotting the concentration of each caffeine standard on the x-axis and the corresponding peak area on the y-axis. The Beer-Lambert Law, where A = abc (refer to Appendix D), states that the relationship between absorbance (A, or peak area) is proportional to the concentration of a substance (c). The resulting straight line (or straight line equation if you prefer to calculate this) will enable you to determine the concentration of caffeine in your sample provided you know the peak area of your sample. A sample calibration curve is shown in Figure 7. This result and the volumes used in diluting the soft drink for analysis will allow you to determine the molarity of caffeine in your soft drink as well as the number of milligrams of caffeine in a 500 mL bottle of your soft drink.

Calculations

Your instructor will provide you with a chromatogram analogous to your chosen soda sample: mountain dew or coke. From this chromatogram identify the peak that corresponds to caffeine (remember to look at a retention time which most closely matches that of your caffeine standards). Determine the peak area of the caffeine peak and record this on your data sheet in the designated Table 3: Soda Sample. Using your graph of peak area vs. caffeine concentration on your data sheet (or the equation of this relationship) to find the value for the molarity of caffeine in your chosen soda sample based on the peak area on the soda sample chromatogram. Record the value of concentration of your chosen soda sample in Table 3 and on the line provided on your data sheet (a). This is the value for M_1 in the following calculation in part (b).

By adding 10 mL of DI water to 10 mL of soda we performed a dilution. To calculate the concentration of caffeine in the original, undiluted soda sample, use Equation 1:

$$M_1 V_1 = M_2 V_2$$

Record this value on the line designated for (b) molarity of caffeine in original soft drink, (M_2). In this calculation M_1 is designated as the molarity of caffeine in the diluted soda sample (determined graphically) and V_1 is the volume of the diluted sample (20.0 mL). V_2 is the volume of the original soda sample (10.0 mL). Solving the equation for M_1 gives the molarity of caffeine in the original soda sample.

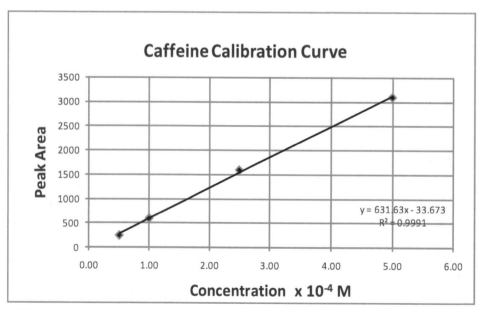

FIGURE 7 — *Example Caffeine Calibration Curve*

(c) To convert the concentration of caffeine from molarity (moles/L) to mg caffeine/500 mL soda, use the following equation 2.

$$M_2 \text{ (mol/L)} \times (194.2 \text{ g/mol caffeine}) \times (1000 \text{ mg/g}) \times (0.500 \text{ L}) = — \text{ mg caffeine in soda bottle} \qquad \text{[Eqn. 2]}$$

Waste Management

All solutions can be poured down the sink. Rinse the pipets, volumetric flasks, and caps with DI water and return to the reagent bench.

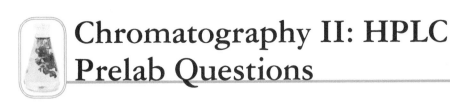

Chromatography II: HPLC
Prelab Questions

Name _____ Section _____

1. Following the directions in the procedure, complete the following table by solving for V_2:

Table 1 *Calculation of Volumes Required to Prepare Standard Caffeine Solutions*

Caffeine Standard (M)	M_1 (Standard Concentration, M)	V_1 (mL)	M_2 (Stock Solution Concentration, M)	V_2 (Volume, mL, to Pipet Stock Solution into Volumetric Flask)
5.00×10^{-4}	5.00×10^{-4}	50.00	5.00×10^{-3}	5.00 mL
2.50×10^{-4}	2.50×10^{-4}	50.00	5.00×10^{-3}	
1.00×10^{-4}	1.00×10^{-4}	50.00	5.00×10^{-3}	
0.500×10^{-4}	0.500×10^{-4}	50.00	5.00×10^{-3}	

2. How many grams of caffeine would be required to prepare a 7.6×10^{-3} M solution in a 250 mL flask (mw of caffeine = 194.2)? Show your calculations.

3. a) Why is the pipet rinsed with soft drink before filling it to the line with soft drink?

   b) How would the final calculated concentration of caffeine in the soda be affected if the meniscus from the soda in the pipet was over the line when the solution was transferred? Why?

Fill line

Chromatography II: HPLC Data Sheet

Name _____ **Section** _____

Name of Lab Partner(s) _____

Table 2 *Information for Standards*

Retention time	Concentration (x-axis)	Peak Area (y-axis)
	0.500×10^{-4} M	
	1.00×10^{-4} M	
	2.50×10^{-4} M	
	5.00×10^{-4} M	

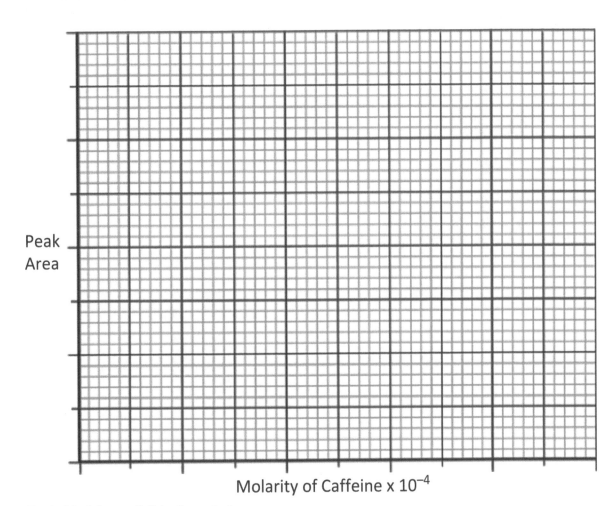

Peak Area

Molarity of Caffeine x 10^{-4}

Graph of Peak Area vs. Caffeine Concentration

Table 3 *Soda Sample:*_____

Retention Time of Caffeine in Soda Sample	Concentration of Caffeine in Soda Sample (x-axis)	Peak Area of Caffeine Peak in Soda Sample (y-axis)

Calculations

a) Molarity of caffeine (M_1) in diluted solution (concentration from x-axis) _____

b) Molarity of caffeine in original soft drink (M_2) _____

c) mg of caffeine in 500 mL of soft drink _____

Chromatography II: HPLC
Review Questions

Name _____ Section _____

1. When the soft drink *Jolt* is used in place of *Coke* in the HPLC experiment, the concentration of caffeine in the injected sample is found to be 5.9×10^{-4} M. What peak area corresponds to this concentration? Explain how you came up with your answer.

2. Explain why caffeine is more soluble in methanol than in water using your knowledge of intermolecular forces.

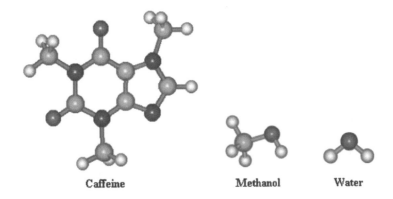

Caffeine   Methanol   Water

3. Caffeine is more soluble in methanol than in water. If you wanted the caffeine to elute from the column more quickly (using the same mobile phase flow rate), how would you change the relative amounts of methanol and water in the mobile phase? Why?

Chromatography II: HPLC
Supplemental Questions

Name _____ Section _____

Colligative Properties: Freezing Point Depression

Introduction

Colligative properties are those properties of a solution that depend only on the *number of solute particles* dissolved in the solution. The greater the concentration of solute in the solution, the greater the effect will be on the colligative properties of the solution. The nature of the solute particles, whether they are atoms, molecules, or ions, is *not* a factor. Examples of colligative properties for solutions include vapor pressure depression, freezing point depression, boiling point elevation, and osmotic pressure elevation.

An example of a colligative property with which you might be familiar is the use of "coolant" in the radiator of your car. A "coolant" is nothing more than a solute (ethylene glycol or propylene glycol) added to the water in your radiator to form a solution. The solution will boil at a higher temperature than pure water (boiling point elevation) so you are less likely to experience radiator "boil over" on a hot day. Of course, this same solution will now freeze at a lower temperature than pure water (freezing point depression), so it also protects your radiator from "freezing up" on a cold day. This is why radiator additives such as Prestone® are often labeled with the seemingly contradictory title of "coolant/antifreeze".

If we dissolve a **molecular substance,** such as sugar ($C_{12}H_{22}O_{11}$), in water, we wind up with as many moles of dissolved particles (sugar molecules) in the solution as there were moles of solid sugar at the beginning.

$$C_{12}H_{22}O_{11(s)} + H_2O_{(l)} \longrightarrow C_{12}H_{22}O_{11(aq)}$$

1 mole of solid   **1 mole** of dissolved molecules

However, if we dissolve an **ionic substance,** such as NaCl, in water, we wind up with *more* moles of dissolved particles (Na^+ ions and Cl^- ions) in the solution than there were moles of solid NaCl at the beginning.

$$NaCl_{(s)} + H_2O_{(l)} \longrightarrow Na^+_{(aq)} + Cl^-_{(aq)}$$

1 mole of solid   **2 moles** of dissolved ions

The number of moles of dissolved particles produced per mole of solute is called the **van't Hoff factor, *i*.** The following table lists examples of compounds and their corresponding *i* values when dissolved in water.

Compound	van't Hoff Factor (i)
All molecular compounds	1
NaCl	2
CaCl$_2$	3
Na$_3$PO$_4$	4

The difference in freezing point between a solution and the pure solvent (ΔT_f) is expressed by the following relationship.

$$\Delta T_f = iK_f m$$

K_f is the **molal freezing point depression constant**[1] for the solvent (K_f for water is 1.86 °C/m), i is the van't Hoff factor for the solute, and **m** is concentration of solute in the solution expressed in terms of **molality**. Molality is defined as

$$m = \frac{\text{moles solute}}{\text{kg solvent}}$$

It is important to use molality instead of molarity in calculations involving colligative properties because, unlike molarity, *the molality of a solution does not change with temperature.*

There are two parts to this experiment. In Part A you will be learning how to prepare solutions using volumetric glassware. In Part B you will be measuring and comparing the freezing point depression of water using the solutions prepared in Part A.

Procedure

This experiment is performed in groups. One pair in the group will be assigned to work with two of the solutes in the following table, while the other pair will work with two different solutes. When the experiment is complete, data can be shared within the group.

Part A: Preparing Solution Concentrations

For each solute assigned to you, calculate the mass (grams) necessary to make the solution concentration specified for that solute in **Table 1** in Prelab Question #2. Record these values in the table under mass of solute (g). Have your instructor review the mass values you have calculated before preparing your solutions.

The mass of NaCl required to prepare a 0.50 M solution of sodium chloride is given as an example using the following calculation:

Molarity of solution × molar mass of solute × volume of solution = grams of solute

Example:

$$0.500 \frac{\text{moles}}{\text{Liter}} \times 58.5 \frac{\text{grams}}{\text{mole}} \times 0.0500 \text{ liter of DI water} = \frac{1.46 \text{ g NaCl}}{\text{to be used}}$$

---

[1]Another name for molal freezing point depression constant is cryoscopic constant.

Obtain a clean and dry 50 mL beaker and record its mass on your data sheet. Into the beaker add the approximate amount of the solute indicated in **Table 1** (see Prelab Question 2). Record the exact mass of the beaker plus solute on your data sheet.

Obtain a 50 mL **volumetric flask** and clean it well with DI water (refer to **Figure 1** below). This unique glassware is made to have a narrow neck to be more precise and is calibrated to contain a specified volume of liquid, such as 50 mL in the volumetric flasks used in today's experiment. In the center of the neck of the volumetric flask is the calibration mark that indicates that specific volume. When preparing a solution, the meniscus of the liquid should be on the calibration mark. Weigh the clean and dried volumetric flask with the cap on the flask, and record the mass on your data sheet.

Weigh your solute in a 50 mL beaker. Add 20 to 30 mL of DI water to the solute in the beaker and swirl to dissolve the solute. Transfer this solution into the volumetric flask using a narrow-necked (small) funnel. Use your wash bottle to rinse the beaker into the volumetric flask and then remove the funnel. Add DI water to the volumetric flask using your wash bottle until the meniscus lies on the calibration mark on the neck of the flask. Place and hold the cap on the volumetric flask while you swirl the contents. Hold the cap tightly and shake the flask until all of the solute is dissolved.

Part B: Freezing Point Measurements

Reweigh the solution in the volumetric flask with the cap on the flask, and record this value on your data sheet. To obtain the mass of the solvent (DI water), subtract the mass of the volumetric flask (including cap) and the mass of the solute from the mass of solution. Convert the solvent mass value into kg. To determine the molality (m) of solution, divide the moles of solute by the kg of solvent.

In today's experiment, you will be using a thermos bottle as an insulated container to measure freezing points of various solutions. Add enough ethanol to the thermos until it is approximately ½ to ¾ full. The ethanol can be reused for subsequent trials.

Using crucible tongs, **carefully** add 3 or 4 small pellets of dry ice to the ethanol in the thermos. ***Do not touch the dry ice with unprotected hands***. Pot holders will be available for you to use to hold the cold thermos bottle. The optimal temperature value for your dry ice bath is –20°C. Open the "Freezing Point Depression CHML 102" experiment under the "Time and Temperature" prompt of the MicroLab data acquisition unit. Use the temperature probe to monitor the temperature of your dry ice bath to ensure it is near the optimum temperature of –20°C. You may need to add or remove dry ice pellets or ethanol to achieve this value. When your dry ice bath is –20°C remove the temperature probe and wipe it clean before using it for your samples.

Volumetric flask ⎯⎯⎯⎯⎯⎯

Calibration mark ⎯⎯⎯⎯⎯⎯

Stirring solute dissolved in 20 or 30 mL of DI water in small beaker before transfer to volumetric flask

FIGURE 1 — *Volumetric Flask in Solution Preparation*

Place only pure DI water into a clean and dry test tube until the test tube is 1/3 full. Place the test tube into the dry ice/ethanol bath in the thermos using your test tube holder. The test tube holder should remain on the test tube so that you can pull it out readily without use of your hands. Put an electronic temperature probe into the DI water in the test tube and click the "Start" prompt of the program while you continuously stir the water with the temperature probe. Monitor the temperature until it levels off near the freezing point of pure water. Once this value is achieved click the "Stop" prompt of the computer program. Record this temperature as $T_{i(water)}$ on your data sheet. This value only needs to be determined once.

Place some of your solution into a clean and dry test tube until it is approximately 1/3 full. Place this test tube into the dry ice/ethanol bath in the thermos using your test tube holder. (Remember to keep the test tube holder on the test tube). Put the temperature probe into the solution in the test tube and continuously stir the solution while monitoring the temperature until the temperature reading is a constant value. Record this temperature value as $T_{f(solution)}$ on your data sheet. Observe how the water from your solution freezes and forms ice crystals near the top and sides of the test tube. Place the test tube in your test tube rack and wait until the solution melts to clean and dry the test tube. Repeat this procedure for the other solute assigned to you.

Data Analysis

The moles of solute are determined by dividing the grams of solute by its molar mass.

$$\text{Moles solute} = \frac{\text{grams solute}}{\text{molar mass}}$$

The molality of each solution is found by dividing the moles of solute by the number of **kilograms of solvent (water)** in the solution.

$$\text{molality} = \frac{\text{moles of solute}}{\text{kg solvent}}$$

The value of ΔT_f is found by subtracting the freezing point of the water from the freezing point of the solution. This should give you a negative number since the freezing point of the solution should be less than the freezing point of the DI water.

$$\Delta T_f = T_{f(solution)} - T_{i(water)}$$

On the back of your data sheet, record the molality (m), van't Hoff factor (i), effective molality (i × m), and your experimentally determined value for ΔT_f. Your experimentally determined value of ΔT_f may be compared to the calculated value for ΔT_f for that solute using the formula:

$$\Delta T_f = i K_f m$$

Safety Precautions

Avoid skin contact with dry ice. If contact occurs, rinse immediately with plenty of water.

Waste Management

When you have completed your trials pour ethanol from the dry ice/ethanol bath into the container labeled "Recyclable Ethanol" located in the fume hood.

All solute solutions may be flushed down the sink with plenty of water.

Colligative Properties: Freezing Point Depression Prelab Questions

Name _____ **Section** _____

1. What determines the value of *i* for a particular solute? Predict the value of *i* for Na_3PO_4. Explain.

2. Following the instructions in the procedure, complete the following chart for mass of solute (g):

Table 1 *Determination of Mass of Solutes (g) to be Used in Preparation of Solutions*

Solute	Molar Mass (g/mol)	Mass of Solute (g)	Volume of Solution (mL)	Concentration (Molarity)
NaCl	58.5	1.46	50.0	0.500
KI	166.0		50.0	0.250
$Mg(NO_3)_2 \cdot 6H_2O$	256.4		50.0	0.500
Sucrose	342.3		50.0	0.500

3. Calculate the mass (g) of $NaCl_{(s)}$ (mw = 58.5) required to dissolve in 250 mL of water to make a 0.300 M solution.

4. If 5.83 g of sucrose (mw = 342.3) are dissolved in enough water to make a 2000 mL of solution, what is the molarity of the solution?

Colligative Properties: Freezing Point Depression Data Sheet

Name _____ **Section** _____

Name of Lab Partner(s) _____

Data

Measurement	NaCl	KI	$Mg(NO_3)_2$	Sucrose
mass of 50 mL beaker				
mass of beaker + solute				
mass of solute				
moles of solute				
mass of 50 mL volumetric flask				
mass of solution + 50 mL vol. flask				
mass of solution − mass of solute				
mass of solvent (kg)				
molality of solution (mol/kg)				
T_i (freezing temp. of water)				
T_f (freezing temp. of solution)				
ΔT_f				

Results

Solute	Molality	van't Hoff factor, i	Effective Molality ($i \times m$)	Experimental ΔT_f	Calculated ΔT_f
NaCl					
KI					
$Mg(NO_3)_2 \cdot 6H_2O$					
Sucrose					

1. Compare the ΔT_f of the NaCl and KI solutions. How do the molality of the solution and the van't Hoff factor (i) of the solute affect ΔT_f? Explain.

2. Compare the ΔT_f of the KI and sucrose solutions. How do the molality of the solution and the van't Hoff factor (i) of the solute affect ΔT_f? Explain.

3. Compare the ΔT_f of the $Mg(NO_3)_2$ and sucrose solutions. Are these results what you would have expected? Explain.

Colligative Properties: Freezing Point Depression Review Questions

Name _____ **Section** _____

1. Arrange the following aqueous solutions in order of <u>decreasing</u> freezing point. (List so the last compound has the lowest freezing point.)

    a) 0.075 molal glucose (a molecular compound)

    b) 0.075 molal LiBr

    c) 0.030 molal $Zn(NO_3)_2$

    d) 0.030 molal NaI

2. Predict the ΔT_f of a 0.50 m solution of $Al(NO_3)_3$. Explain.

3. The freezing point of 1.0 liter of water was determined to be $-3.5°C$. How many grams of NaCl have been added to the water? (Assume that the density of water is 1.0 g/mL.)

Name _____ **Section** _____

Equilibrium Constant, K$_{eq}$

Introduction

Most chemical reactions do not go to completion. During the course of most reactions, a state is reached in which the rate of the forward reaction is equal to the rate of the reverse reaction. The reaction is said to be in a state of **dynamic equilibrium**. When a chemical reaction has reached equilibrium, the forward and reverse reactions are still taking place, but *there is no net change in the concentrations of reactants or products*. Since the reactant and product concentrations are not changing, the ratio of the concentration of products to the concentration of reactants must be a constant. This is known as the **equilibrium constant, K$_{eq}$**.

Consider the following equation representing a chemical reaction in which reactants A and B form products C and D.

$$a\,A + b\,B \rightleftharpoons c\,C + d\,D$$

The double arrows ($\rightleftharpoons$) indicate that this reaction is one that reaches an equilibrium state. A mathematical expression can be written that represents the constant ratio of products to reactants at equilibrium. This is known as the **equilibrium expression**.

$$K_{eq} = \frac{[C]^c[D]^d}{[A]^a[B]^b}$$

Brackets are used to represent the equilibrium concentrations of reactants and products. The only factor that affects the equilibrium constant for a chemical reaction is temperature. The specific effect of temperature on K$_{eq}$ depends on the sign of ΔH for the reaction.

The objective of today's experiment is to measure the value of the equilibrium constant for the chemical reaction shown below.

$$Fe^{3+}_{(aq)} + SCN^-_{(aq)} \rightleftharpoons FeSCN^{2+}_{(aq)}$$
pale yellow   colorless   orange-red

The equilibrium expression for this reaction is

$$K_{eq} = \frac{[FeSCN^{2+}]}{[Fe^{3+}][SCN^-]}$$

You will determine the value of K$_{eq}$ by mixing together known concentrations of Fe(NO$_3$)$_3$ (iron(III) nitrate) and KSCN (potassium thiocyanate). By measuring the concentration of FeSCN^{2+} present at equilibrium, you will be able to calculate the equilibrium concentrations of all the species shown in the equilibrium expression above.

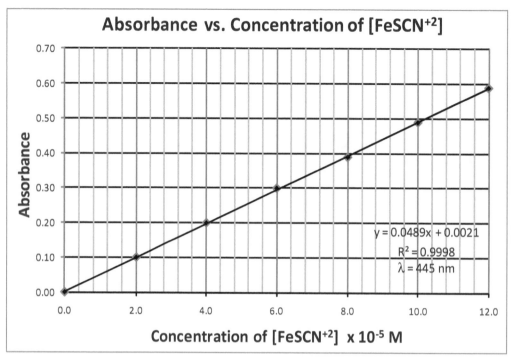

FIGURE 1 — *Absorbance vs. [FeSCN²⁺]*

A spectrophotometer will be used to determine the equilibrium concentration of $FeSCN^{2+}$ by measuring the amount of light absorbed by the orange-red colored complex ion. Measuring concentrations with a spectrophotometer is quick and it does not disturb the chemical system being studied. Your lab instructor will demonstrate the procedure for measuring the absorbance of solutions using the MicroLab spectrophotometer. You should refer to **Appendix D** for a more detailed description of the theory of spectrophotometry, the Beer-Lambert Law, and instructions for using the MicroLab spectrophotometer.

As part of this experiment, you will create a calibration curve by measuring the absorbance of a set of solutions that contain <u>known concentrations</u> of three solutions of $FeSCN^{2+}_{(aq)}$. An example of a calibration curve is shown in Figure 1.

The calibration curve will then be used to convert the absorbance measurements you make on your other five solutions directly into concentration values for $FeSCN^{2+}_{(aq)}$. Instructions for determining the equilibrium concentrations of reactants from the [FeSCN²⁺] data are detailed below in the Calculations section.

Procedure

Part A: Generating a Calibration Curve

Setup your laboratory computer and the MicroLab data acquisition system. Open the MicroLab software and select Spectrophotometer Experiment. Put 10 mL of DI water in a cuvette to serve as the blank for the experiment. Obtain a set of three standards which contain <u>known concentrations</u> of $FeSCN^{2+}_{(aq)}$ from the reagent shelf and fill 3 corresponding **cuvettes** approximately ¾ full of each. Cuvettes are specialized glassware of uniform thickness and clarity designed for making spectrophotometric measurements. Place lids on each of these 4 cuvettes and keep a record of standards placed in each without writing on the cuvette lids. You will generate a calibration curve using these standards solutions of $FeSCN^{2+}_{(aq)}$ before you analyze your samples from trials 1–5.

a) Using a Kimwipe, gently wipe the outside of each cuvette to remove any residual fingerprints. Place the blank in the cuvette holder of the MicroLab spectrophotometer and cover it with the black film canister. The program has already initiated the "#1 Calibrate" setting. Click the "Read Blank" prompt.

b) Once the blank absorbance value has been determined the program is ready to analyze standards in the "#2 Read". The default setting of the spectrophotometer is "% Transmittance". Click the "Absorbance" setting on the top menu to switch settings. Remove the blank from the cuvette holder and place the standard with the lowest concentration of $FeSCN^{2+}_{(aq)}$ in the spectrophotometer. Press the "Add" button. In the dialog box that appears, enter "Std 1" for the Sample ID and enter the actual concentration of the standard in the Concentration box using the "e" for exponent rather than "× 10" (for example: $1.8e^{-5}$). Press the "OK" button.

c) In the Spectrum Profile in the upper right hand corner of the program, click on the blue bar that represents wavelength at 470 nm (or select the closest wavelength to 470 nm).

d) Repeat step (b) for the remaining standards labeling them "Std 2", "Std 3". In today's experiment we only have 3 standards, but other experiments may use several more. Also, note that it is not necessary to select the 470 nm, or closest wavelength again, but you may want to learn how to choose another, more optimal wavelength by highlighting any other you wish. Note that when you do choose another wavelength, the absorbance values change accordingly.

e) Click on the menu tab labeled "#3 Curve" and select the box labeled "Linear" for the curve fit. The computer will automatically generate your calibration curve. Record the concentrations of the standard solutions of $FeSCN^{2+}_{(aq)}$ and their corresponding absorbance values on the first page of your data sheet under "Generating a Calibration Curve". Note that units of absorbance are arbitrary and can be labeled "a.u." for "absorbance units". Record the equation of the line, correlation coefficient and wavelength for your calibration curve on your data sheet.

f) Leave the computer program open, in which the calibration curve is displayed, while preparing your solutions for trials 1–5. Continue with the following procedure to make your unknown solutions.

Part B: Preparation and Analysis of Samples from Trials 1 through 5

Label five clean, dry, regular test tubes 1 to 5, or note their positions in your test tube rack. Pour about 30 mL of the 2.00×10^{-3} M $Fe(NO_3)_3$ solution into a dry 100 mL beaker. Pipet 5.0 mL of this solution into each test tube using a 5 mL volumetric pipet.

Add about 20 mL of the 2.00×10^{-3} M KSCN solution to another dry 100 mL beaker. Using a 10 mL transfer pipet, add 1, 2, 3, 4 and 5 mL from the KSCN beaker into each of the corresponding test tubes as indicated in the Table 1.

Finally, using a transfer pipet cleaned well with DI water, pipet the appropriate amount of DI water into each of the test tubes to bring the total volume in each of the test tubes to 10.0 mL (see Table 1). Mix each solution thoroughly with a glass stirring rod, being sure to rinse and dry the rod after mixing each solution. Pour each solution into a separate cuvette, making sure that you keep track of which solution is in each cuvette. Do not write on the cuvette lids.

Place a cuvette representing Trial 1 in the spectrophotometer. Click on the tab labeled "4 Read" in the upper menu bar. Press the "Add" button and enter "Trial 1" for the Sample ID. (The prompt box will not ask concentration because this is not known.) Press the "OK" button. The computer will now display the equilibrium concentration of $FeSCN^{2+}_{(aq)}$ of sample for trial 1 in the table on the left hand side of the computer program, and the sample will be displayed on the calibration curve by a white "×". Record the absorbance and concentration for $[FeSCN^{+2}]_{eq}$ for trial 1 on your data sheet under graphical analysis.

Repeat the above step for the remaining four trials. Record the absorbance and corresponding concentration values for $[FeSCN^{+2}]_{eq}$ for each trial on your data sheet. Follow the instructions in the Calculations section to determine the value of K_{eq} for each trial.

Safety Precautions

Handle solutions with care as they are dissolved in HNO_3 which may cause burns and holes in clothing.

Waste Management

All wastes may be flushed down the sink with tap water.

Calculations

(Trial 1 has been performed as an example.)

Step 1. *Initial Number of Moles of Each Reactant*

For each trial, you can calculate the number of moles of each reactant from the initial concentrations and volumes listed in Table 1.

Table 1 *Concentrations and Volumes of Reactants and Water*

Trial	2.00×10^{-3} M $Fe(NO_3)_3$	2.00×10^{-3} M KSCN	H_2O
1	5.0 mL	1.0 mL	4.0 mL
2	5.0 mL	2.0 mL	3.0 mL
3	5.0 mL	3.0 mL	2.0 mL
4	5.0 mL	4.0 mL	1.0 mL
5	5.0 mL	5.0 mL	0.0 mL

initial moles of reactant = molarity × volume = (moles/liter) × liter

initial moles Fe^{3+} = $(2.00 \times 10^{-3}$ moles/L$) \times (5.0 \times 10^{-3}$ L$)$ = 1.0×10^{-5} mol Fe^{3+}

initial moles SCN^- = $(2.00 \times 10^{-3}$ moles/L$) \times (1.0 \times 10^{-3}$ L$)$ = 2.0×10^{-6} mol SCN^-

The *initial* concentration of the product, $FeSCN^{2+}$, is zero.

	Fe^{3+}	SCN^-	$FeSCN^{2+}$
Initial moles	1.0×10^{-5} mol	2.0×10^{-6} mol	0 mol
Change in moles			
Equilibrium moles			

Step 2. Determining Concentration of Product Formed and Conversion to Number of Moles of Product

The concentration of the product, $[FeSCN+2]_{eq}$, is determined experimentally by measuring the absorbance of the solution and then converting this to concentration using the calibration curve you will create. For example, your sample for trial 1 has an absorbance reading of 0.09 and, according to your calibration curve, this corresponds to the $FeSCN^{2+}$ concentration of 1.8×10^{-5} mol/L. These values will be recorded on your data sheet under "graphical analysis" as in the example below:

TRIAL 1

Graphical analysis: Absorbance of $FeSCN^{2+}$ = __0.09 a.u.__ $[FeSCN^{2+}]_{eq}$ = __$1.8 \times 10^{-5} M$__

	Fe^{3+}	SCN^-	$FeSCN^{2+}$
Initial moles	1.0×10^{-5} mol	2.0×10^{-6} mol	0 mol
Change in moles			
Equilibrium moles			

The value of concentration of $[FeSCN^{2+}]_{eq}$ assumes use of 1 L of solution. However, because this experiment uses only 10 mL in each trial, the concentration of $[FeSCN^{2+}]_{eq}$ must be converted from moles/L to moles in order to determine the equilibrium amounts of Fe^{3+} and SCN^-. Follow the procedure below to calculate the number of moles of $FeSCN^{2+}$ from concentration of $FeSCN^{2+}$ (in M):

equilibrium moles of $FeSCN^{2+}$ = molarity × volume = (moles/liter) × liter

Note:

The total volume of the solution containing the $FeSCN^{2+}$ is 10 mL, which is 0.010 L or 10×10^{-3} L.

equilibrium moles $FeSCN^{2+}$ = $(1.8 \times 10^{-5}$ mol/L $FeSCN^{2+}) \times (10 \times 10^{-3}$ L) = 1.8×10^{-7} mol $FeSCN^{2+}$

	Fe^{3+}	SCN^-	$FeSCN^{2+}$
Initial moles	1.0×10^{-5} mol	2.0×10^{-6} mol	0 mol
Change in moles			
Equilibrium moles			1.8×10^{-7} mol

Step 3. Change in number of Moles of Each Reactant

According to the stoichiometry of the reaction, one mole of Fe^{3+} and one mole of SCN^- are used up for every mole of $FeSCN^{2+}$ formed. Therefore, the change in the number of moles of each reactant is equal to the moles of product formed at equilibrium:

change in Fe^{3+} = 1.8×10^{-7} mol Fe^{3+}

change in SCN^- = 1.8×10^{-7} mol SCN^-

	Fe^{3+}	SCN$^-$	FeSCN^{2+}
Initial moles	1.0×10^{-5} mol	2.0×10^{-6} mol	0 mol
Change in moles	1.8×10^{-7} mol	1.8×10^{-7} mol	1.8×10^{-7} mol
Equilibrium moles			1.8×10^{-7} mol

Because reactant molecules are used up in the reaction, a negative sign is put in front of their change. Similarly, because FeSCN^{2+} is formed in the reaction, a positive sign is put in front of its change.

	Fe^{3+}	SCN$^-$	FeSCN^{2+}
Initial moles	1.0×10^{-5} mol	2.0×10^{-6} mol	0 mol
Change in moles	$- 1.8 \times 10^{-7}$ mol	$- 1.8 \times 10^{-7}$ mol	$+ 1.8 \times 10^{-7}$ mol
Equilibrium moles			1.8×10^{-7} mol

Step 4. Moles of Reactants and Product at Equilibrium

The moles of reactants present at equilibrium can be determined by adding the change in moles to the initial number of moles.

moles of reactant at equilibrium = Initial Number of Moles of reactant + change in moles of reactant

equilibrium moles Fe^{3+} = 1.0×10^{-5} mol + (-1.8×10^{-7} mol) = 9.8×10^{-6} mol Fe^{3+}

equilibrium moles SCN$^-$ = 2.0×10^{-6} mol + (-1.8×10^{-7} mol) = 1.8×10^{-6} mol SCN$^-$

equilibrium moles FeSCN^{2+} = 0 + 1.8×10^{-7} mol = 1.8×10^{-7} mol FeSCN^{2+}

	Fe^{3+}	SCN$^-$	FeSCN^{2+}
Initial moles	1.0×10^{-5} mol	2.0×10^{-6} mol	0 mol
Change in moles	$- 1.8 \times 10^{-7}$ mol	$- 1.8 \times 10^{-7}$ mol	$+ 1.8 \times 10^{-7}$ mol
Equilibrium moles	9.8×10^{-6} mol	1.8×10^{-6} mol	1.8×10^{-7} mol

Step 5. Concentration of Reactants and Product at Equilibrium

The concentration of FeSCN^{+2} is obtained directly using the spectrophotometer in the lab. The equilibrium concentrations of the reactants, Fe^{3+} and SCN$^-$, can be determined by taking the equilibrium number of moles and dividing by the volume of solution, keeping in mind that the equilibrium volume is the total volume after mixing all reactants (this is 10 mL, which is 0.010 L or 10×10^{-3} L, for all trials).

$$\text{molarity of reactants at equilibrium} = \frac{\text{equilibrium moles of reactant}}{\text{total volume}}$$

molarity of Fe^{3+} at equilibrium $= \dfrac{9.8 \times 10^{-6}\ \text{mol}}{10 \times 10^{-3}\ \text{L}} = 9.8 \times 10^{-4}\ \text{mol/L}$

molarity of SCN^- at equilibrium $= \dfrac{1.8 \times 10^{-6}\ \text{mol}}{10 \times 10^{-3}\ \text{L}} = 1.8 \times 10^{-4}\ \text{mol/L}$

Remember, the concentration of $FeSCN^{2+}$ for this trial was determined experimentally and was found to be $1.8 \times 10^{-5}\ \text{mol/L}$.

Step 6. Calculating K_{eq} for the Reaction

Once the equilibrium concentrations are known, you can substitute them directly into the equilibrium expression to obtain the value of the equilibrium constant.

$$K_{eq} = \frac{[FeSCN^{2+}]}{[Fe^{3+}][SCN^-]}$$

$$K_{eq} = \frac{[1.8 \times 10^{-5}\ \text{mol/L}]}{[9.8 \times 10^{-4}\ \text{mol/L}][1.8 \times 10^{-4}\ \text{mol/L}]} = 102$$

Step 7. Calculating the Average K_{eq} for the Reaction

When data is collected for all 5 trials, an average K_{eq} value may be calculated. Record this value on your data sheet.

Equilibrium Constant, K$_{eq}$
Prelab Questions

Name _____ **Section** _____

1. Which one of the following statements about a chemical reaction at equilibrium is correct?

   a) The concentrations of the reactants and products are equal.

   b) The rates of the forward and reverse reactions are equal.

   c) No more reactants are converted into products.

   d) The concentrations of reactants and products change with time.

2. What compound is analyzed using the spectrophotometer? Explain the main principle behind this method of analysis (Hint: use **Appendix D**).

3. Using the graph of absorbance vs. [FeSCN^{2+}] in Figure 1, what is the concentration of a FeSCN^{2+} solution which has an absorbance of 0.4?

Equilibrium Constant, K$_{eq}$ Data Sheet

Name _____ **Section** _____

Name of Lab Partner(s) _____

Part A: Generating a Calibration Curve

Wavelength (nm) = _____

Standards of FeSCN$^2_{(aq)}$:

      Standard #1 concentration = _____

                    absorbance = _____

      Standard #2 concentration = _____

                    absorbance = _____

      Standard #3 concentration = _____

                    absorbance = _____

Equation of line (y = ax + b) = _____

    The term y refers to _____

    The term x refers to _____

  Correlation coefficient = _____

Part B: Analysis of Samples in Trials 1 through 5

Trial 1

Graphical analysis:

Absorbance of $FeSCN^{2+}$ = _____ Concentration of $[FeSCN^{2+}]_{eq}$ = _____ M

	Fe^{3+}	SCN^-	$FeSCN^{2+}$
Initial moles	1.0×10^{-5} mol	2.0×10^{-6} mol	0 mol
Change in moles	mol	mol	mol
Equilibrium moles	mol	mol	mol

× 0.010 L

÷ 0.010 L   ÷ 0.010 L

Concentration of $[Fe^{3+}]_{eq}$ = _____ M $[SCN^-]_{eq}$ = _____ M

$$K_{eq} \text{ for Trial 1} = \frac{[FeSCN^{2+}]_{eq}}{[Fe^{3+}]_{eq}[SCN^-]_{eq}} = \underline{\hspace{2cm}}$$

Trial 2

Graphical analysis: Absorbance of $FeSCN^{2+}$ = _____ $[FeSCN^{2+}]_{eq}$ = _____

	Fe^{3+}	SCN^-	$FeSCN^{2+}$
Initial moles			
Change in moles			
Equilibrium moles			

Concentration of $[Fe^{3+}]_{eq}$ = _____ $[SCN^-]_{eq}$ = _____

$$K_{eq} \text{ for Trial 2} = \frac{[FeSCN^{2+}]_{eq}}{[Fe^{3+}]_{eq}[SCN^-]_{eq}} = \underline{\hspace{2cm}}$$

Trial 3

Graphical analysis: Absorbance of $FeSCN^{2+}$ = _____ $[FeSCN^{2+}]_{eq}$ = _____

	Fe^{3+}	SCN^-	$FeSCN^{2+}$
Initial moles			
Change in moles			
Equilibrium moles			

Concentration of $[Fe^{3+}]_{eq}$ = _____ $[SCN^-]_{eq}$ = _____

K_{eq} for Trial 3 = _____

Trial 4

Graphical analysis: Absorbance of $FeSCN^{2+}$ = _____ $[FeSCN^{2+}]_{eq}$ = _____

	Fe^{3+}	SCN^-	$FeSCN^{2+}$
Initial moles			
Change in moles			
Equilibrium moles			

Concentration of $[Fe^{3+}]_{eq}$ = _____ $[SCN^-]_{eq}$ = _____

K_{eq} for Trial 4 = _____

Trial 5

Graphical analysis: Absorbance of $FeSCN^{2+}$ = _____ $[FeSCN^{2+}]_{eq}$ = _____

	Fe^{3+}	SCN^-	$FeSCN^{2+}$
Initial moles			
Change in moles			
Equilibrium moles			

Concentration of $[Fe^{3+}]_{eq}$ = _____ $[SCN^-]_{eq}$ = _____

K_{eq} for Trial 5 = _____

Average K_{eq} of 5 Trials = _____

Equilibrium Constant, K$_{eq}$
Review Questions

Name _____ **Section** _____

1. What was the effect on the equilibrium constant of increasing the concentration of KSCN in the reaction?

2. What effect did the increasing concentration of KSCN have on the concentration of the product?

3. In your own words, explain what your average K$_{eq}$ value means in terms of the reaction between $Fe^{+3}_{(aq)}$ and $SCN^{-}_{(aq)}$ to form $FeSCN^{+2}_{(aq)}$. (Is the reaction reactant or product favored?)

4. The equilibrium constant, K$_{eq}$, for the following reaction was determined to be 184.

$$Fe^{+3}_{(aq)} + SCN^{-}_{(aq)} \rightleftharpoons FeSCN^{+2}_{(aq)}$$

At equilibrium, the concentration of the $FeSCN^{+2}$ ion was measured spectro-photometrically to be 7.83×10^{-5} M. If the equilibrium concentration of Fe^{+3} was determined to be 9.3×10^{-4} M, what was the equilibrium concentration of the thiocyanate ion?

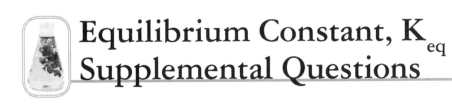

Equilibrium Constant, K$_{eq}$
Supplemental Questions

Name _____ **Section** _____

Exchange Reactions

Introduction

In this experiment you will have the opportunity to observe many **exchange reactions** (also called **metathesis reactions, precipitation reactions, or double displacement reactions**) and to write balanced equations to represent these reactions. Deductive reasoning will be used to identify a set of unknown solutions based on the products formed through systematic combinations of each of your unknowns.

An exchange reaction is one that involves the exchange of cations and anions between reactants. When one of the exchange reaction products is insoluble in the mixture, a **precipitate** (abbreviated ppt) forms. This insoluble product is usually observed as a cloudiness or turbidity in the mixture, a gelatinous suspension, or solid particles settling to the bottom. The exchange reactions you will observe during this experiment yield solid products that settle to the bottom of the reaction mixture. If no solid precipitate is formed, then "NR" is written to indicate "no reaction".

In an exchange reaction, the cations and anions of the two reactants "switch partners" as shown in the equation below.

$$A\textbf{B} + C\textbf{D} \longrightarrow A\textbf{D} + C\textbf{B}$$

Here is an example of an exchange reaction involving barium chloride and sodium sulfate.

$$BaCl_{2\,(aq)} + Na_2SO_{4\,(aq)} \longrightarrow BaSO_{4\,(s)} + 2\,NaCl_{(aq)}$$
$$\text{white ppt.}$$

The barium sulfate ($BaSO_4$) formed in this reaction precipitates as a white solid and collects at the bottom of the test tube. The NaCl remains dissolved in the solution.

The ability of $BaCl_2$ and Na_2SO_4 to undergo an exchange reaction and form the insoluble, white $BaSO_4$ can be used as a means of confirming or disproving the presence of either $BaCl_2$ or Na_2SO_4 in a solution. For example, if a solution is suspected of containing Na_2SO_4, the addition of $BaCl_2$ will produce a white precipitate if Na_2SO_4 is present. If no precipitate is formed, then Na_2SO_4 cannot be present in the solution.

In today's experiment you will be issued a piece of paper on which are written the formulas for ten compounds and on the reverse side the numbers of your five unknowns. Your task is to identify the five compounds from your list of ten possibilities. To aid you in your identifications, a table of insoluble compounds, rules for determining solubility, and certain other properties of selected compounds are included in the lab instructions.

Procedure

Step 1. Obtain a list of possibilities and an unknown number set from your instructor. Copies both lists and return the slip of paper to your lab instructor. Do not mark on this paper.

During the course of the experiment, it is extremely important that all glassware (test tubes, droppers, stirring rods, disposable pipets, etc.) be rinsed thoroughly with deionized water (labeled DI on your faucet) <u>before and after</u> each test.

Obtain your unknowns *one at a time* by using the pipet attached to the unknown bottle to deliver approximately 3 mL (1/6 of a test tube) to a clean test tube. *Be sure to number your unknowns correctly and return the pipet to the correct unknown bottle.*

Step 2. Observe the color of each unknown. The color of a solution can signal the presence or absence of particular ions. For example, if a solution is blue, it may contain the Cu^{2+} ion. If the solution is colorless, it does not contain the Cu^{2+} ion.

Step 3. In order to determine the odor of an unknown, hold the test tube about a foot from your nose and gently fan the vapors toward your nose. This procedure is called **wafting**. There are only two compounds in this experiment that have an odor. Ammonium hydroxide, NH_4OH, will give off vapors that smell like household cleaners that contain ammonia. Ammonium sulfide, $(NH_4)_2S$, has a very offensive odor similar to the smell of rotten eggs. Your instructor will inform you of the unknowns that contain ammonium sulfide and these will be located in one of the fume hoods. When you are ready to perform exchange reactions with ammonium sulfide, take your spot plate *into the hood and perform each reaction one at a time to avoid sulfide vapor contaminating your other reactions*. (This also prevents ammonium sulfide from permeating the air in the laboratory.) When your reactions with ammonium sulfide are complete, pour the wastes directly into the "Sulfides" waste container located in the same fume hood. Use the wash bottle labeled for sulfides waste to rinse out all the sulfide precipitates from your spot plate into the "Sulfides" waste container.

Step 4. Determine the acidity or basicity of each unknown by placing a small quantity of the solution on a strip of litmus paper using your disposable pipet. This allows you to get more use out of one piece of litmus paper since many samples can be placed on the same strip. Do <u>not</u> dip litmus paper directly into your unknowns. This will contaminate your unknowns and possibly cause them to change color. Be careful in interpreting the results of litmus tests on colored solutions. Some solutions, such as those containing transition metal ions, are only weakly acidic and may cause blue litmus to turn a very pale pink color. You should rely on other tests to positively identify such substances.

$$Red\ litmus \longrightarrow Blue = BASE$$
$$Blue\ Litmus \longrightarrow Red = ACID$$

Step 5. In the 5 × 5 reaction matrix on your data sheet, you will record the colors of the precipitates formed when you mix your unknowns with one another, two at a time. Write your unknown numbers along the top and down the side of the matrix. Place two drops of one of the unknowns in a spot plate well. Now place two drops of one of the remaining unknowns in that same well. It is important to avoid contamination by adding drops of unknown solutions to the spot plate wells using clean disposable pipets. Rinse a stirring rod in DI water and stir the mixture in the well. Record all of your observations in the appropriate square of the reaction matrix. If no precipitate forms, write NR to indicate no reaction. If a precipitate forms, write the color of the precipitate in the matrix. Repeat the above procedure with all of your other unknowns until all combinations have been tested. Refer to your list of possibilities to determine what your unknowns may be.

Additional Tests. Some of the unknowns may require you to perform a few additional tests in order to definitively identify them. For example, you may have two blue solutions, which indicates that they may contain copper(II) ions. The exchange reactions performed in your reaction matrix may not be sufficient to distinguish between the two copper compounds. If you encounter this problem, you instructor will give you a solution of $BaCl_2$ that will allow you to distinguish between $Cu(NO_3)_2$ and $CuSO_4$.

Step 6. When you have identified your five unknowns, write balanced equations on your data sheet for all the exchange reactions that occurred during your tests. Indicate the precipitate in each reaction and record the color of the precipitate underneath. Do not write equations for reactions that did not produce a precipitate. The completed matrix, balanced equations, and unknown identities should be returned to your instructor.

Safety Precautions

Avoid directly inhaling vapors of ammonium hydroxide (ammonia) and ammonium sulfide. These substances are strong irritants of the mucous membranes of your respiratory system. If you suspect that one or more of your unknowns contain these substances, smell them using the wafting technique described in Step 3 of the Procedure.

Ammonium sulfide $(NH_4)_2S$ has very offensive and potentially toxic vapors. All tests with ammonium sulfide should be performed only in the fume hood and all sulfide wastes should be deposited into the "Sulfides" waste container in the fume hood.

Waste Management

Pour all waste from this experiment into the appropriately labeled container in the fume hood. Waste containers will be provided for "Chromate", "Sulfide" and general "Metals" wastes. If you are unsure of the identity of your unknowns and precipitates, deposit them in the "Metals" waste container. Do not pour unknowns from your test tubes back into the original containers, as this may introduce contamination.

Chemical Properties of Selected Compounds

The following list contains examples of compounds that are acidic or basic in aqueous solution.

Acidic: HCl, HNO_3, $FeCl_3$, $Hg(NO_3)_2$, NH_4Cl

Basic: $NaOH$, NH_4OH, K_2CrO_4, Na_2CO_3, Na_3PO_4, Na_2SO_3, $(NH_4)_2S$

The following list contains examples of compounds that exhibit **colors** when in solution:

$Cu(NO_3)_2$, $CuSO_4$ - blue

$FeCl_3$ - orange-yellow

K_2CrO_4 - bright yellow

$Ni(NO_3)_2$, $NiSO_4$ - green

$(NH_4)_2S$ - yellow

$CoSO_4$ - red

The following compounds evolve easily identifiable **odors** when in solution:

$(NH_4)_2S$ - rotten eggs *and* household cleaners that contain ammonia

NH_4OH - household cleaners that contain ammonia

Solubility Rules

1. Compounds of the Group IA metals and NH_4^+ are *soluble*.

2. Binary compounds of Cl, Br, and I are *soluble*, except for compounds formed between these nonmetals and the ions of silver(I), mercury(I), lead(II), and copper(I).

3. Compounds containing the NO_3^- group are all *soluble*.

4. Compounds containing SO_4^{2-} are *soluble*, except for the sulfates of strontium, barium, and lead(II).

5. Binary compounds of sulfur are *insoluble*, except for the sulfides of Groups IA and IIA.

6. Compounds containing the OH^- group are *insoluble*, except for the hydroxides of Group IA and strontium and radium.

7. Compounds of metals containing CO_3^{2-} and PO_4^{3-} are *insoluble*, except for those of Group IA compounds.

Common Insoluble Compounds

Bromides (Br⁻)

Ag^+—pale yellow; curdy

Hg_2^{++}—white

Pb^{++}—white

Chlorides (Cl⁻)

Ag^+—white; curdy

Pb^{++}—white; sol. in hot water

Hg_2^{++}—white

Chromates (CrO₄²⁻)

Ag^+—dark orange

Ba^{++}—light yellow

Cu^{++}—orange-brown

Hg^{++}—orange

Pb^{++}—bright yellow

Zn^{++}—bright yellow

Iodides (I⁻)

Ag^+—pale yellow; curdy

Hg^{++}—bright orange

Pb^{++}—bright yellow

Cu^{++}—orange brown

Carbonates (CO₃²⁻)

Ag^+—yellow-white (or gray-white)

Ba^{++}—white

Ca^{++}—white

Cu^{++}—light blue

Hg^{++}—dark orange

Ni^{++}—pale green; gelatinous

Pb^{++}—white

Zn^{++}—white; gelatinous

Hydroxides (OH⁻)

Ag^+—gray-brown

Ba^{++}—white

Ca^{++}—white

Cu^{++}—w/NaOH—pale blue

w/NH₄OH—dark blue solution

Fe^{+++}—rust colored

Hg^{++}—white to yellow

Ni^{++}—Pale green; gelatinous

Pb^{++}—white

Zn^{++}—white; gelatinous

Phosphates (PO₄³⁻)

Ag^+—light yellow

Ba^{++}—white

Ca^{++}—white

Hg^{++}—white

Pb^{++}—white

Zn^{++}—white

Sulfates (SO₄²⁻)

Ba^{++}—white

Pb^{++}—white

Sulfides (S²⁻)

Ag^+—brown

Cu^{++}—brown to black

Fe^{+++}—black

Hg^{++}—black

Ni^{++}—black

Pb^{++}—black

Zn^{++}—white

Exchange Reactions Prelab Questions

Name _____ **Section** _____

1. Write complete balanced equations for the following reactions in aqueous solution. Identify any precipitates that would form and list their color. If no reaction would occur, write NR. You may refer to the table of solubility rules provided with this experiment to determine which product is the precipitate.

   $Pb(NO_3)_2$ + KI ⟶

   $Ca(NO_3)_2$ + $NaCl$ ⟶

   $AgNO_3$ + $K_2Cr\text{-}O_4$ ⟶

   $CuSO_4$ + $Pb(NO_3)_2$ ⟶

2. Write formulas for the following reactants and write the balanced exchange reaction that would occur upon mixing these reactants together.

   ammonium carbonate and barium chloride

   iron(II) phosphate and silver(I) nitrate

   ammonium phosphate and calcium chloride

3. Circle the compounds that are soluble in water:

$NaNO_3$	$(NH_4)_2SO_4$	$CaCO_3$
KCl	Ag_2S	$LiBr$
PbI_2	$PbCl_2$	
$(NH_4)_2S$	$AgNO_3$	

Exchange Reactions Data Sheet

Name _____ **Section** _____

Name of Lab Partner(s) _____

Unknown
Numbers: _____ _____ _____ _____ _____

Color
_____ _____ _____ _____ _____

Odor
_____ _____ _____ _____ _____

Acid/Base/
Neutral _____ _____ _____ _____ _____

Identity of
Unknown _____ _____ _____ _____ _____

Reaction Matrix

Unknown Numbers ⟶					
	X				
	X	X			
	X	X	X		
	X	X	X	X	
	X	X	X	X	X

111

Balanced Exchange Equations

1. Write complete balanced equations for all exchange reactions in which your unknown combinations produced a precipitate. Be sure to identify and list the color of any precipitate formed.

Exchange Reactions Review Questions

Name _____ **Section** _____

1. Write the balanced equation for the reaction of barium nitrate with sodium chromate. Is this an exchange reaction? If so, be sure to indicate the precipitate.

2. Write the balanced equation for the reaction of copper(II) nitrate with potassium bromide. Is this an exchange reaction? If so, be sure to indicate the precipitate.

3. A student is given an unknown to identify. The unknown is colorless, odorless, and neutral. It produces no precipitates when reacted with any of the three halides, Cl^-, Br^-, and I^-. It produces a yellow precipitate when added to sodium chromate. Which of the following could it be? Write its exchange reaction with sodium chromate.

   HCl   NaOH NH_4OH $BaCl_2$ K_2CrO_4 $AgNO_3$ $NiSO_4$

4. A second student in the lab is given the same list of possibilities as the student in question 3 above. This unknown, while also colorless, odorless and neutral, gave a brownish precipitate when mixed with NaOH, and a white precipitate when added to $BaCl_2$. Reactions with the other halides gave yellow precipitates. What unknown do they have? Write the equation for the exchange reaction with the unknown and NaOH. Also, write the balanced reaction between the unknown and $BaCl_2$.

Exchange Reactions Supplemental Questions

Name _____ **Section** _____

Fermentation and Distillation

Introduction

In this experiment you will prepare ethanol (ethyl alcohol), C_2H_5OH, by the fermentation of sugar by yeast. Ethanol is the main ingredient of alcoholic beverages. During **fermentation**, enzymes in the yeast organisms decompose the sugar into carbon dioxide, water, ethanol, and a large number of other substances. When the ethanol concentration in the mixture reaches about 12%, fermentation stops because yeast cells cannot survive in alcohol concentrations above 12%.

The ethanol in the fermentation mixture will be separated from water by **distillation**. Distillation is a physical separation process based on the boiling points of liquids in a mixture. The lower the boiling point of a compound, the more **volatile** it is, and the easier it will be to separate from a mixture, provided the difference in boiling points between the compounds is sufficiently great. As a mixture of two or more liquids is heated, the lower boiling liquid will start to boil and its vapors will separate from the liquid mixture. By cooling the hot vapors, it is possible to condense the vapors back to the liquid state and collect them. This liquid is called the **distillate** and is a relatively pure solution of the lower boiling liquid. In this experiment, ethanol is the lower boiling compound and will be collected for analysis.

After the alcohol has been prepared and distilled, its **proof** will be determined by measuring its density. Proof is defined as two times the volume percent of alcohol in the mixture. You will also determine the proof of some commercial bourbon. This will be done by collecting mass and volume data for bourbon from each pair of students in the lab. These data will be plotted and the proof determined by graphical analysis.

Procedure

Week 1: Fermentation

During the first week of the experiment you will prepare the alcohol. The distillation will be performed during the second week.

To a 250 mL flask add 20 g of sugar (sucrose) that has been weighed in a small beaker, 85 mL deionized (DI) water, and 10 mL of Pasteur salts (a mixture of inorganic nutrients developed by Louis Pasteur while he worked for the French wine industry). Swirl the flask to dissolve all of the solids. Using a small clean beaker, weigh 1.0 g of dry active yeast. Add the yeast to the mixture and gently swirl the flask for 30 seconds.

Add about 8 mL of $Ca(OH)_2$ solution to a medium test tube. Calcium hydroxide reacts with carbon dioxide (CO_2), a product of yeast fermentation, to form a white solid called calcium carbonate. The formation of calcium carbonate in the test tube is a good indication that fermentation has occurred. The following reactions explain the production of calcium carbonate in the $Ca(OH)_2$ layer in the test tube.

$$CO_2 + H_2O \longrightarrow H_2CO_3 \text{ (carbonic acid)}$$
$$H_2CO_3 + Ca(OH)_2 \longrightarrow CaCO_3 \text{ (white solid)} + 2 H_2O$$

Carefully add 4–5 drops of paraffin oil to the test tube so it forms a layer on top of the $Ca(OH)_2$ solution. Support this test tube in a 125 mL flask. Finish the assembly of the fermentation apparatus as shown in Figure 1. It is very important that the rubber tubing remain in the $Ca(OH)_2$ solution throughout the fermentation process to prevent evaporation of the liquids in the flask and test tube. The paraffin oil acts as an extra barrier and seals off the fermentation system. A piece of paper towel is used as a wedge to keep the tubing from working its way out of the test tube. In addition, the oil prevents exposure of surplus oxygen to the fermentation system, which could convert the alcohol to acetic acid (vinegar).

When fermentation is detected, as evidenced by the formation of bubbles in the $Ca(OH)_2$ solution, the complete assembly should be labeled and saved, undisturbed, until the next lab period.

Week 2: Distillation

PART A: LAB ALCOHOL

The following steps are performed during the second week of the experiment. Remove the rubber stopper and tubing. Rinse the tubing with water to remove the white solid adhering to it and return to your lab instructor. Discard the $Ca(OH)_2$/oil solution in the waste beaker located in the fume hood. Avoid shaking the flask and disturbing the sediment. Carefully **decant** the fermentation mixture into a clean 250 mL flask to remove solid matter. To decant a solution, you gently pour off the top liquid layer without stirring up the sediment. Decantation is a technique used to separate a heterogeneous mixture of solid and liquid in which the solid, being denser than the liquid, settles to the bottom.

Assemble the distillation apparatus shown in Figure 2. The collection test tube should be well secured by a clamp attached to the ring stand. In your set-up, be certain that the glass tubing does not touch the bottom of the collecting test tube, which would create a closed system and would cause your distillate to flow in reverse as soon as the heat is cut off. The glass tubing should be assembled so that it is at half the depth of the collecting test tube. In addition, expose as much surface area of the collection test tube with the ice water as possible in order to condense your distillate effectively. Ask your Instructor to check your apparatus if you are uncertain.

Slowly heat the flask until the contents begin to boil very gently. Control the heat to keep the mixture boiling smoothly. You should be able to see vapor moving through the glass tubing and condensing in the test tube. Continue heating until you have collected about 6 mL of distillate (approximately 1/3 test tube full), then remove the glass tubing from the test tube.

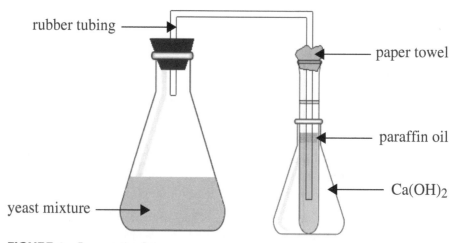

FIGURE 1 — *Fermentation Set-up*

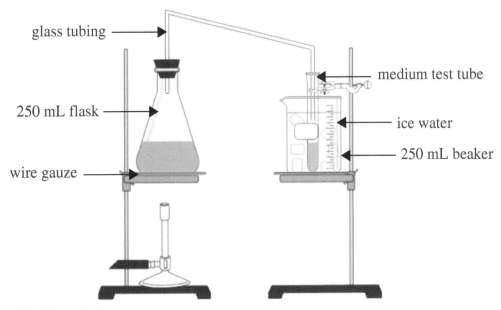

FIGURE 2 — *Distillation Set-up*

Labels in figure: glass tubing, 250 mL flask, wire gauze, medium test tube, ice water, 250 mL beaker

Weigh a clean and dry 10 mL graduated cylinder. Pour approximately 4 mL of your distillate into the graduated cylinder and weigh it again. The density of the alcohol is found by dividing the mass of the alcohol by its volume. The proof of the alcohol is read from the graph of Proof vs. Density, Figure 4, on your data sheet. Indicate the proof of your alcohol with a **labeled line** on Figure 4.

As an additional rough determination of the purity of your lab alcohol, place 10 drops of the alcohol you made on a watch glass. Make certain that everything flammable is clear of your work area before igniting the distillate sample. Attempt to light the alcohol with your Bunsen burner and observe the color of the flame. Alcohols burn with a characteristic blue, nearly invisible flame. Any water in your distillate will be recognized by yellow and orange fringes to the flame of your burning distillate sample. Blow out the flame or place your watch glass under the water faucet to extinguish the flame. Record your observations on the data sheet.

PART B: BOURBON

Each pair of students will be assigned a volume of bourbon for which you are to determine the corresponding mass. The results should be recorded in the table of volume versus mass on the whiteboard in the lab. When the data for every group is on the board, copy it to your data sheet. You will determine the density and proof of the alcohol in bourbon by using this data and graphical analysis. Because this is an important procedure used in several experiments, a summary follows.

Graphical Analysis

Graphical analysis is often employed to find the mathematical relationship between two variables. In this experiment, the relationship between the mass of a substance and its volume is studied.

When data for two variables (we will call these x and y for now) are collected in the laboratory, the data can be plotted together on a graph using **regular** intervals for the axes. The values are plotted using the *entire* graph ("bunching" of data increases the probability of error). The **independent variable** is plotted on the horizontal axis (x-axis) and the **dependent variable** on the vertical axis (y-axis). A graph of this sort is called a **scatter plot.**

It is not always easy to determine which variable in an experiment is the independent variable and which is the dependent variable. In today's experiment, the mass of your bourbon sample *depended* on the volume of the sample you were given, so mass is the dependent variable (plotted on the y-axis) and volume the independent variable (plotted on the x-axis).

Many of the relationships we study among variables in General Chemistry are linear ones. That means that once the x and y data are plotted, a "best-fit" straight line may be drawn (with a ruler, not free hand) through the data points. A "best fit" line is one that minimizes the distance of all the data points from the line.

The general formula for a straight line is $y = mx + b$, where x is the independent variable, y is the dependent variable, m is the slope of the line, and b is the y-intercept (the value of y when x is zero). If the line goes through the origin (the point 0,0), then b = 0. The equation then becomes $y = mx$, which is referred to as a *direct* relationship. The formula for determining the slope of a straight line from the graph is:

$$m = \frac{\Delta y}{\Delta x} = \frac{y_2 - y_1}{x_2 - x_1}$$

The coordinates chosen for determining the slope should be points on the best-fit straight line, not any of the data points you plotted.

Calculate the slope of the straight line shown in Figure 3. Mark the two sets of coordinates you chose by placing an "X" on the line for each one. Does this line represent a direct relationship between x and y?

Now consider the data you obtained from the class for the mass and volume of various samples of bourbon. A plot of mass vs. volume for the bourbon samples is a straight-line relationship whose slope is the density of bourbon. You can use the density of the bourbon to calculate its proof using Figure 4, which can be found on the Data Sheet at the end of this experiment. Indicate the proof of your bourbon with a *labeled line* on Figure 4.

Safety Precautions

<u>**Do not drink**</u> the alcohol you produced in this experiment. Unknown contaminants from the glassware could become mixed with the alcohol.

Waste Management

Discard the $Ca(OH)_2$ and oil solution from the fermentation set-up in your lab sink and flush with plenty of water. Wash the test tube well with soap and water. The alcohol residues can be flushed down the drain with plenty of water.

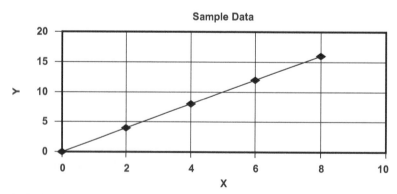

FIGURE 3— *Scatter Plot with Best-Fit Line*

Fermentation and Distillation
Prelab Questions

Name _____ **Section** _____

1. Distillation separates components of a mixture based on their:

2. Explain the term "decant".

3. Calculate the slope of the graph in Figure 3 of the Graphical Analysis section of the Procedure. Record the chosen coordinates and slope value in the space below.

Fermentation and Distillation Data Sheet

Name _____ **Section** _____

Name of Lab Partner(s) _____

Part A: Lab Alcohol

1. Mass of graduated cylinder + alcohol   _____

2. Mass of empty graduated cylinder   _____

3. Mass of alcohol   _____

4. Volume of alcohol   _____

5. Density of lab alcohol   _____

6. Proof of lab alcohol
    (from Figure 4)   _____

7. In the space below, record your observations when attempting to light the alcohol you made.

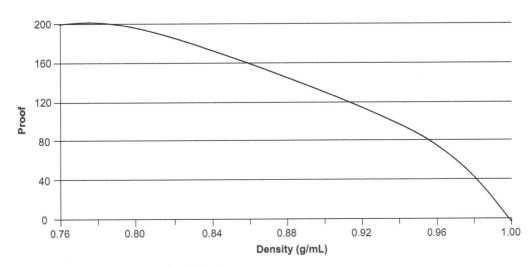

FIGURE 4—*Proof vs. Density for Alcohol*

Part B: Bourbon Class Results

Mass (g)	Volume (mL)

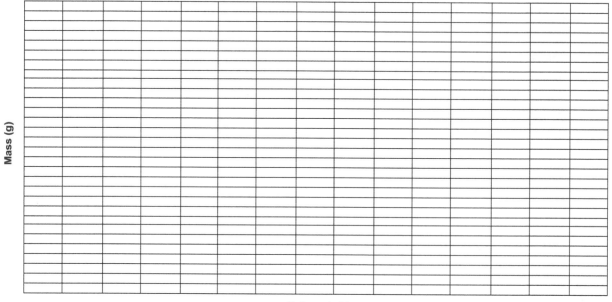

Mass (g)

Volume (mL)

Calculation of Slope

(Show work from graphical analysis)

Density of ABC bourbon (from slope) _____

Proof of ABC bourbon (from Figure 4) _____

Fermentation and Distillation Review Questions

Name _____ **Section** _____

1. What were two observed pieces of evidence that fermentation was occurring?

2. Predict what would happen to your distillate if you distilled two substances of equivalent boiling points.

3. What physical process occurs in the tubing that is connected between the flask being heated (the alcohol/water mixture) and the collecting test tube (the distillate)? Is this a chemical or physical process?

Fermentation and Distillation Supplemental Questions

Name _____ **Section** _____

Formulas and Equations

Introduction

One of the important attributes of humans is the ability to communicate complex ideas to one another using language. The branch of science called chemistry has its own unique language, which consists of **chemical formulas** and **chemical equations**. These are the means by which chemists represent substances and the changes substances undergo. A review of these concepts follows.

Chemical Formulas

Formulas are used to represent the number and kind of atoms making up a basic unit of a **compound**. A compound is a substance resulting in the combination of elements held together by chemical bonds. Symbols are used to denote the atoms in the compound. Compounds that contain metals combined with nonmetals are said to be **ionic**. Compounds composed of nonmetals and nonmetals are **molecular** compounds. All of the compounds in today's experiment are ionic. The formula for zinc sulfide, ZnS, represents one zinc metal ion combined with one nonmetal sulfide ion. The formula of an ionic compound represents the smallest whole number ratio of positive ions, called **cations,** to negative ions, called **anions**. In this example, the symbol for zinc as a cation is accompanied by the appropriate positive charge, Zn^{2+}. Zinc is **monatomic** because the ion is formed from only one atom. Sulfide is written as a negatively charged anion, S^{2-}, and is also monatomic.

The chemical formula of an ionic compound is always written so that the sum of the charges on positive and negative ions equals zero. Zinc has a 2+ charge and sulfide has a 2- charge. The correctly written formula for zinc sulfide is ZnS because this is the smallest number of zinc ions and sulfide ions for which the charges cancel. Chemical formulas always convey the assembly of atoms in fixed proportions which are denoted by the subscript associated with each ion. The subscript "1" is understood for both Zn and S ions in the compound ZnS, and is not written.

Consider ammonium, NH_4^+, which is an example of a **polyatomic** ion. It consists of five atoms (one nitrogen and four hydrogen atoms) combined to form a *single ion unit*. When ammonium is combined with sulfide, S^{2-}, the compound is written as $(NH_4)_2S$, which indicates two ammonium ions combined with one sulfide ion. This is the smallest whole number ratio of these ions for which all of the charges cancel. Note that the formulas for compounds consist of atoms in fixed proportions with subscripts used to balance the charges of individual ions.

Transition metals often form more than one positive ion. Therefore, the positive charge of transition metal ions is indicated by a Roman numeral in parentheses immediately following the ion's name. For example, Co^{2+} is written as cobalt(II) while Co^{3+} is written as cobalt(III). The names of the corresponding cobalt chloride compounds are written as cobalt(II) chloride and cobalt(III) chloride. Although lead (Pb) and tin (Sn) are not true transition metals, Roman numeral designations are necessary because they commonly form ions with more than one charge.

Table 1 shows ions that will be used in today's laboratory exercise and in the pre-and post-laboratory questions.

Table 1 *Common Ions and Their Charges*

	Cations (+)			Anions (–)	
Monatomic	**Polyatomic**		**Monatomic**	**Polyatomic**	
1+	**1+**		**1 –**	**1 –**	
Hydrogen, H^+			Bromide, Br^-	Acetate, $C_2H_3O_2^-$ (CH_3COO^-)	
Cesium, Cs^+	Ammonium, NH_4^+		Chloride, Cl^-	Bicarbonate, HCO_3^-	
Copper(I), Cu^+			Fluoride, F^-	Chlorate, ClO_3^-	
Lithium, Li^+			Iodide, I^-	Cyanide, CN^-	
Mercury(I), Hg_2^{2+}				Hydrogen sulfate, HSO_4^-	
Potassium, K^+				Hydroxide, OH^-	
Silver(I), Ag^+				Nitrate, NO_3^-	
Sodium, Na^+				Nitrite, NO_2^-	
				Perchlorate, ClO_4^-	
				Permanganate, MnO_4^-	
				Thiocyante, SCN^-	
2+	**2+**		**2-**	**2-**	
Barium, Ba^{2+}			Oxide, O^{2-}	Carbonate, CO_3^{2-}	
Beryllium, Be^{2+}			Peroxide, O_2^{2-}	Chromate, CrO_4^{2-}	
Cadmium(II), Cd^{2+}			Sulfide, S^{2-}	Dichromate, $Cr_2O_7^{2-}$	
Calcium, Ca^{2+}				Hydrogen Phosphate, HPO_4^{2-}	
Chromium(II), Cr^{2+}					
Cobalt(II), Co^{2+}				Sulfate, SO_4^{2-}	
Copper(II), Cu^{2+}				Sulfite, SO_3^{2-}	
Iron(II), Fe^{2+}					
Lead(II), Pb^{2+}					
Magnesium, Mg^{2+}					
Manganese(II), Mn^{2+}					
Mercury(II), Hg^{2+}					
Nickel(II), Ni^{2+}					
Strontium, Sr^{2+}					
Tin(II), Sn^{2+}					
Zinc(II), Zn^{2+}					
3+	**3+**		**3 –**	**3 –**	
Aluminum, Al^{3+}			Nitride, N^{3-}	Arsenate, AsO_4^{3-}	
Cobalt(III), Co^{3+}				Ferricyanide, $Fe(CN)_6^{3-}$	
Chromium(III), Cr^{3+}				Phosphate, PO_4^{3-}	
Iron(III), Fe^{3+}					

Chemical Equations

Chemical equations are written summaries of the changes that are thought to take place during a chemical reaction. During a chemical reaction, the atoms of the **reactants** (starting materials) combine and rearrange to form new chemical substances. These new substances are the **products** of the chemical reaction. Evidence that a chemical reaction has occurred includes such things as color change, temperature change, formation of a precipitate, and evolution of a gas.

A chemical equation is usually written to show reactants on the left side of the arrow and products on the right. The arrow indicates the direction in which the reaction occurs. The equation must be **balanced,** that is it must be written to show equal numbers and kinds of atoms on the left as there are on the right. **Coefficients** are numbers that appear in front of a formula in an equation and are used to balance the equation. Your instructor will review the basics of balancing chemical equations.

Procedure

This experiment is performed by each student alone. It is important that you learn and understand the material covered in these exercises. The concepts presented here are essential in the execution and understanding of future experiments.

Part A: Observations

On the left end of the reagent shelf you will find six labeled bottles containing 0.1M solutions of ionic compounds dissolved in water. This is Set 1. Observe the color of each solution and record your observations in the appropriate spaces on the data sheet. Some ions exhibit characteristic colors when dissolved in water. These colors are so stable that they can be used to identify the presence (or absence) of a particular ion in a solution.

Special Note: When describing solutions and heterogeneous mixtures it is important to distinguish between the terms clear, cloudy, colored, and colorless. A glass of strawberry Kool-Aid is *clear* and *red*, while a glass of tomato juice is *cloudy* and *red*.

From the names on the bottles and information found in the Introduction, you are to write the formulas for each compound. The first entry in the table has already been completed as an example. Do not remove the solutions from the shelf.

When you have completed Set 1, have your instructor review the formulas you have written. Once you have written all six correctly, your lab instructor will initial your data sheet. Do not proceed to the next step until your formulas have been approved.

The next table on the data sheet asks you to determine the color exhibited by each ion in the solutions you observed. You will need to combine the results of all your observations to answer this correctly.

In the middle of the reagent shelf are six more solutions. This is Set 2. Follow the same procedure with Set 2 as you did with Set 1. It is not necessary for your instructor to approve the formulas you write for this set before you go to the next step.

At the right end of the reagent shelf are four solutions comprising Set 3. Proceed with this set as you did with Set 2.

Set 4 consists of four unknown solutions. You are to observe these solutions and predict which ions could be responsible for the colors you observe. Do not remove these solutions from their location.

Part B: Spot Plate Reactions

In this part of the exercise you will carry out three chemical reactions and write balanced equations for these reactions. The reactions are performed by adding two or three drops of each of two reactants to a *spot plate*. The spot plate allows you to conveniently perform micro-scale reactions, thus conserving chemical materials and reducing the amount of chemical waste.

Write the balanced equations including the correct formulas for each reactant and product. Identify the product that forms a precipitate in each mixture by using the designation "(s)" to denote a solid, and record the color of the precipitate. For help in determining which of the products is the precipitate, refer to the solubility rules given by your instructor. Remember, nitrates are <u>always</u> soluble; therefore the product which is not a nitrate is the solid precipitate. The first equation is already completed as an example.

Finally, you are to use the *Handbook of Chemistry and Physics,* published by the Chemical Rubber Company (CRC), to obtain information such as color and melting point in order to correctly confirm each precipitate. **Please do not write or mark in the *Handbook.***

Waste Management

The products from spot plate reactions #2 and #3 of Part B of this experiment should be rinsed into the "Chromates waste" container in the fume hood using a wash bottle.

Rinse the products of spot plate reaction #4 of Part B into the "Metals waste" container in the fume hood using a wash bottle.

Formulas and Equations
Prelab Questions

Name _____ Section _____

1. Give the correct chemical formula for the following (you may refer to Table 1 in the Introduction):

    iron(III) oxide   barium chloride

    lead(II) chloride   chromium(III) nitrate

    ammonium carbonate   iron(II) phosphate

    sodium acetate   magnesium phosphate

    calcium hydroxide   cobalt(II) nitrate

    potassium chromate   mercury(II) iodide

    tin(II) chloride   manganese(II) oxide

2. Name each of the following ionic compounds:

    a) $Ca(CH_3COO)_2$

    b) $Co_2(SO_4)_3$

    c) $Al(OH)_3$

3. Which of the following are the correct formulas of compounds? For those that are not, give the correct formula.

    a) Ca_2O

    b) $SrCl_2$

    c) K_2S

Formulas and Equations Data Sheet

Name _____ **Section** _____

Name of Lab Partner(s) _____

Part A

Set 1

Compound Name	Formula	Color in Solution
1. sodium nitrate	$NaNO_3$	colorless
2.		
3.		
4.		
5.		
6.		

Instructor's initials _____

Ion	Color of Ion in Solution
1. Na^+	
2. K^+	
3. Cl^-	
4. NO_3^-	
5. CrO_4^{2-}	

Set 2

Compound Name	Formula	Color in Solution
1. ammonium sulfate	$(NH_4)_2SO_4$	colorless
2.		
3.		
4.		
5.		
6.		

Ion	Color of Ion in Solution
1. SO_4^{2-}	
2. Cu^{2+}	
3. NH_4^+	
4. Ni^{2+}	
5. Fe^{3+}	
6. Ag^+	

Set 3

Compound Name	Formula	Color of Solution
1.		
2.		
3.		
4.		

Ion	Color of Ion in Solution
1. Pb^{2+}	
2. Co^{2+}	
3. Mg^{2+}	
4. Ca^{2+}	

Unknown Number	Color of Solution	Ions Responsible for Color
I		
II		
III		
IV		

Part B

1. silver(I) nitrate + sodium chloride ⟶ silver(I) chloride + sodium nitrate

$$AgNO_3 + NaCl \longrightarrow AgCl_{(s)} + NaNO_3$$
$$\text{(white)}$$

2. silver(I) nitrate + potassium chromate ⟶

3. lead(II) nitrate + potassium chromate ⟶

4. lead(II) nitrate + sodium sulfate ⟶

Properties of Precipitates from <u>Handbook of Chemistry and Physics:</u>

Name of Precipitate	Crystalline Form	Melting Point	Page Number & Edition
silver(I) chloride	White, cubic	455 °C	B-219 46th ed.

Formulas and Equations
Review Questions

Name _____ **Section** _____

1. Predict the colors of the following compounds in aqueous solution.

Compound	Color in Solution
1. $CoSO_4$	
2. $Cu(NO_3)_2$	
3. $(NH_4)_2CrO_4$	
4. NH_4NO_3	
5. $FeCl_3$	
6. Ag_2SO_4	
7. $NiSO_4$	
8. KNO_3	
9. $Pb(NO_3)_2$	

2. Write the complete and balanced equation for the reaction of $Ba(NO_3)_2$ and Na_2CrO_4. Identify the precipitate and record its color.

3. An unknown aqueous solution is discovered in the lab. The solution is colorless. Which of the following can be ruled out as <u>not</u> being present? Which could possibly be present?

$AgNO_3$ $CuSO_4$ NH_4Cl $Ni(NO_3)_2$ Na_2CrO_4 $NaCl$ HCl

4. Write the balanced equations including the correct formulas for each reactant and product.

a) zinc(II) chloride + lead(II) nitrate ⟶

b) calcium iodide + copper(II) sulfate ⟶

c) manganese(II) sulfate + barium nitrate ⟶

Formulas and Equations
Supplemental Questions

Name _____ **Section** _____

Fractional Crystallization

Introduction

Many of the common substances around us are made of complex **heterogeneous mixtures** of elements or compounds. A mixture is defined as a substance in which two or more materials are combined which do not react chemically. The individual compounds or elements in a mixture remain pure and unchanged, maintaining their unique properties. It is often necessary to separate a mixture into its components in order to gain a better understanding of these substances, or it may be necessary to separate and remove them as impurities in order to purify a substance. The objective of this experiment is to introduce you to one of the more commonly used methods for separating mixtures.

Recall that each individual substance in a mixture retains its unique set of physical properties. The separation procedure described in this experiment takes advantage of differences in the solubilities of substances in an aqueous (water) solution. Changing the temperature of the solution can increase the differences in solubilities of the solutes in the solution. Simply stated, some of the components of the mixture are more soluble in hot water than in cold water.

Substances that remain insoluble in water as undissolved solids can be removed from the mixture by **filtration**. Filtration is the process of physically separating a solid from a liquid in which the solid remains partly or completely in suspension. A porous material such as a filter is used to allow only liquid to pass through. The liquid material, including all substances dissolved in it, is called the **filtrate**. Pay close attention to the experimental procedure to determine the necessity for saving or removing the filtrate.

A Büchner funnel attached to an Erlenmeyer flask with an attached side arm (side-arm flask) acts as a vacuum filtration system. The side arm is connected to rubber hosing which is fitted to an aspirator on the faucet. The filtration process is aided by suction caused by reduced air pressure from tap water passing rapidly through the faucet. Filter paper placed in the funnel allows the **residue**, or solid material, to be trapped and removed from the filtrate. The residue can be analyzed further to determine the identity and quantity of its components.

In this experiment, we will take advantage of the physical property of solubility differences to separate a mixture of sand (SiO_2), sodium chloride (NaCl), and borax (sodium tetraborate decahydrate, $Na_2B_4O_7 \cdot 10H_2O$). This separation method is known as **fractional crystallization**.

Sand is insoluble in water and sodium chloride is very soluble in both hot and cold water. Borax however, is quite soluble in hot water, but is relatively insoluble in cold water. The mixture can be **fractionated**, or separated, by adding hot water to dissolve the NaCl and Borax, leaving insoluble sand behind to be removed by filtration. The filtrate is heated to evaporate some of the water and then cooled to crystallize the borax. The borax crystals are then fractionated from the NaCl, which is dissolved in the cool water, by vacuum filtration.

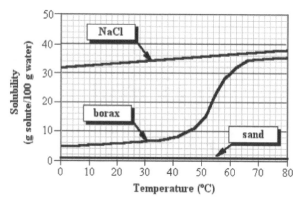

FIGURE 1 — *Solubility versus Temperature*

Figure 1 shows a graph of solubility versus temperature for each of the substances we will separate from the mixture in today's experiment. The graph clearly shows the solubility differences between borax and NaCl at low temperatures.

Procedure

Weigh a 250 mL beaker. Select one of the fractional crystallization mixtures and add approximately 25 g of it to the 250 mL beaker you just weighed. Reweigh the beaker with the mixture to determine the weight of the mixture. Add 60 mL DI (deionized) water to the beaker.

Place the mixture in the 250 mL beaker on a wire gauze on a ring stand and heat the mixture to 85°C as in the photo in Figure 2. Do not allow the mixture to boil. DO NOT STIR WITH YOUR THERMOMETER; rather, use the glass stirring rod to stir the solution. Use the thermometer for periodically monitoring the temperature only. Stir the mixture for five minutes to dissolve all soluble materials.

While the mixture is heating, construct the suction filtration apparatus shown in Figure 3. Weigh the filter paper and record the value on your data sheet. *(Special note: never heat the suction flask in the burner flame).* The Büchner funnel apparatus separates the heterogeneous mixture containing solid and liquid phases. For best results, the water faucet containing the aspirator should be turned completely on and the filter paper moistened with a small amount of deionized water before filtering.

After the mixture in the 250 mL beaker has been heated and stirred for five minutes, turn on the aspirator and pour the mixture into the Büchner funnel. Be sure to filter the mixture while it is still hot. Use your rubber policeman to transfer as much of the sand as possible to the filter paper. You should now have some sand on the filter paper and 60 mL of clear, colorless filtrate in the flask. *Break the vacuum by removing the hose from the aspirator, then turn off the water.* If you turn off the water before breaking the vacuum, some water may wash back into the flask possibly contaminating the filtrate.

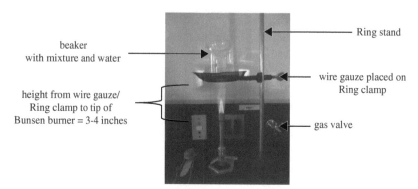

FIGURE 2 — *Heating Fractional Crystallization Mixture in Water using the Bunsen Burner*

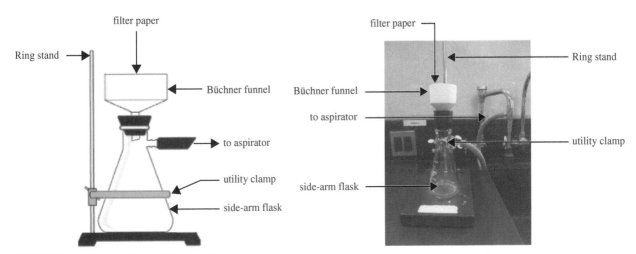

FIGURE 3—*Büchner Funnel Filtration Set-up*

Pour the filtrate into a clean 250 mL beaker and save it for the collection of the borax. Wash the sand in the Büchner funnel with several portions of DI (deionized) water while applying suction to the flask. Continue to apply suction to aid in drying the sand. Discard the wash water that collects in the flask. Weigh a watch glass and record this value. Remove the filter paper containing the sand from the funnel using your spatula and place the sand residue on the pre-weighed watch glass to dry. The residue may be weighed and the weight of the watch glass and filter paper can be subtracted after they are dry.

Heat the solution in the 250 mL beaker to boiling and allow it to gently boil for three minutes. This will reduce the volume of water in the solution. If the solution bumps or spatters while being heated, cover the beaker with a watch glass and reduce the heat slightly. After the solution has boiled for three minutes, cool to room temperature (approximately 25°C) in a water bath by placing tap water in a larger beaker. Once the solution reaches room temperature, it can be further cooled by adding ice to the tap water in the larger beaker. White crystals of borax should form in the beaker as it cools in the ice bath. Figure 1 illustrates that borax becomes insoluble when the temperature is 25°C and lower. If crystals have not formed after ten minutes, scratch the inside bottom of the beaker with a glass stirring rod. The scratches cause small chips of glass to be removed from the stirring rod and will hasten the crystallization process by providing a nucleus around which crystals can form.

Separate the white borax crystals from the mixture using the suction filtration apparatus. Pre-weigh another piece of filter paper and record this value. Place the pre-weighed filter paper in your Büchner funnel and wet it down with a small amount of DI water so that it is flush against the bottom of the funnel. Pour the mixture containing the borax crystals into the Büchner funnel. Use your rubber policeman to transfer the last remaining borax crystals to the Büchner funnel. While applying suction, wash the crystals by pouring 10 mL of ice-cold deionized water over them. Allow the suction to continue for several minutes to help dry the crystals by removing most of the water. Turn off the aspirator and remove the filter paper containing the borax crystals. Using your spatula, place the borax residue and filter paper on another pre-weighed watch glass. Weigh the borax and filter paper residue. Subtract the weight of the filter paper and watch glass when recording the weight of the borax. Discard the filtrate in the flask.

According to Figure 1, it is not possible to recover all the borax in your mixture, as 8 g of borax will always remain dissolved in 100 g of water between 0°C and 10°C. However, in calculating the % borax in your sample, assume you obtained 100% recovery. In other words, we will assume that all of the borax remained insoluble in the cold water.

Both NaCl and borax are white crystals. How do you know that you collected borax instead of NaCl in the Büchner funnel? Look up the melting points of sodium chloride and sodium tetraborate decahydrate (borax) in the *Handbook of Chemistry and Physics* (section: Physical Constants of Inorganic Compounds). Record these values on the data sheet. Refer to Figure 4, which illustrates the approximate temperatures of the main regions of a Bunsen burner flame and determine the part(s) of the Bunsen burner flame in which your white borax crystals would melt. How could you differentiate between the two salts based on their melting locations in the flame?

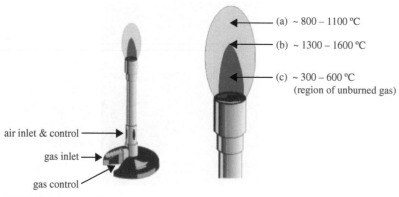

(a) ~ 800 – 1100 °C

(b) ~ 1300 – 1600 °C

(c) ~ 300 – 600 °C
(region of unburned gas)

air inlet & control

gas inlet

gas control

FIGURE 4— *The Bunsen Burner*

Safety Precautions

Your instructor will approve your fractional crystallization set-up before you supply heat using the Bunsen burner. Remove and transfer hot solutions using pot holders to grasp and secure beakers. *Do not* use crucible tongs or test tube holders for transferring hot solutions. The Bunsen burner, ring clamp and wire gauze will remain hot for a period of time after heating. Please use caution when touching these for disassembly.

Waste Management

Clean the Büchner funnel and side-arm flask thoroughly. Be sure to return the rubber tubing, Büchner funnel and side-arm flask to the lab shelves. Clean your balance area well. After you complete the experiment, place your recovered sand and borax into the garbage container.

Fractional Crystallization
Prelab Questions

Name _____ Section _____

1. What physical property is utilized for the separation of the mixture into its separate components? How is this property controlled?

2. Answer the following questions, which refer to Figure 1 in the Introduction section of this experiment:

   a) Which substance shown on the graph is the most soluble in water at 60°C?

   b) At 25°C, how much of the borax remains dissolved in 100 g of water?

   c) What temperature is necessary to dissolve 30 g of borax in 100 g of water?

   d) How much of a 45 g sample of NaCl would dissolve in 100 g of water at room temperature? How much of the 45 g of NaCl would remain undissolved?

   e) You are given a solid mixture containing 20 g of NaCl and 30 g of borax. This mixture completely dissolves in water at 70°C. Now, suppose the solution is allowed to cool to 10°C. You observe white crystals of borax form in the flask. What happened to the NaCl?

Fractional Crystallization Data Sheet

Name _____ **Section** _____

Name of Lab Partner(s) _____

1. Weight of sample mixture and 250 mL beaker   _____

2. Weight of 250 mL beaker   _____

3. Weight of sample mixture (about 25 g)   _____

4. Weight of sand, filter paper and watch glass   _____

5. Weight of filter paper   _____

6. Weight of watch glass   _____

7. Weight of sand   _____

8. Percent of sand in mixture   _____

9. Weight of recrystallized borax, paper, and watch glass   _____

10. Weight of filter paper   _____

11. Weight of watch glass   _____

12. Weight of recrystallized borax   _____

13. Percent of borax in mixture   _____

14. Weight of NaCl in sample (by difference)   _____

15. Percent of NaCl in mixture   _____

16. Melting points of NaCl and borax from
    Handbook of Chemistry and Physics   _____

17. Region(s) of flame in which borax would melt   _____

18. Region(s) of flame in which NaCl would **not** melt   _____

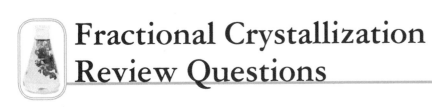

Fractional Crystallization
Review Questions

1. How will the calculated percent of NaCl in you unknown be affected if not all of the borax and sand were recovered. Explain.

2. A mixture contained 2.11 g SiO_2, 0.23 g NH_4Cl, and 7.66 g $CaCO_3$. What is the percent of $CaCO_3$ in the sample?

3. A, B and C have the following solubilities in 100 mL of water at 25 °C, respectively: 15 g, 35 g and 65 g. If 75 g of each are added to 200 mL water, which will completely dissolve? How much of each will remain undissolved?

Name _____ **Section** _____

Gas Laws

Introduction

In this experiment, you will study the behavior of a gas (air)[1] as you vary its **pressure, volume,** and **temperature.** By doing so, you will be able to determine the relationship that exists between pressure and volume as well as the relationship between pressure and temperature for a gas. Of the three states of matter (solid, liquid, and gas), gases are the easiest to study. Unlike the **condensed states** of matter (solids and liquids), the properties of gases are completely described by only four variables; pressure (P), temperature (T), volume (V), and the number of moles of gas (n). For example, if you know the pressure, temperature, and volume of a sample of gas, it is possible to calculate the number of molecules of gas in the sample.

Over the years, scientists have developed a theory to help explain why gases behave the way they do. This theory, called the **Kinetic Molecular Theory**, has four major postulates.

1. Gases are composed of atoms or molecules whose size is much smaller than the distance between them. Therefore, we assume that the volume of gas molecules is negligible.

2. Gas molecules move randomly at various speeds in all directions.

3. Except when they collide, gas molecules exert *no forces on other gas molecules in the system*. All collisions are elastic.

4. The average kinetic energy (KE) of the gas molecules in a sample is proportional to the temperature of the gas.

The nanoscale model of gas behavior based on these assumptions predicts the macroscopic behavior of many gases at room temperature and atmospheric pressure very accurately. An **ideal gas** is one which obeys <u>all</u> of the postulates of the Kinetic Molecular Theory under <u>all</u> conditions of pressure and temperature. While some **real gases** exhibit nearly ideal gas behavior under certain conditions, all real gases deviate from ideal gas behavior at low temperatures and high pressures. For example, if you cool a sample of nitrogen gas below 77 K (−196 °C), it will liquefy. Air, the gas used in today's experiment, behaves nearly ideally at room temperature and atmospheric pressure.

The pressure and temperature measurements for this experiment will be performed using a computer and data acquisition equipment, which will graph your data in real-time. Your lab instructor will demonstrate the proper procedure for using the software and the temperature and pressure probes.

---

[1]Air is actually a homogeneous mixture of primarily two gases, O_2 and N_2, but as with all gas mixtures, it behaves as though it were a single substance.

Procedure

Part A: Boyle's Law: Pressure vs. Volume

In the first part of the experiment, you will be adjusting the **volume** of a syringe to study the effect on the **pressure** of the gas in the syringe at constant temperature. Set up your lab computer and MicroLab unit and start the data acquisition software by opening the MicroLab software. Click on the Gas Laws tab and then open the Boyles Law Experiment template. Obtain a plastic syringe and adjust the plunger to the 10 mL mark. Connect the syringe to the pressure port on the back of the MicroLab unit using a length of plastic tubing. The pressure reading on the computer should be approximately 760 torr (760 torr = 760 mmHg = 1 atm pressure). If the reading is not close to 760 torr, ask your instructor for assistance.

Press the "Start" button to begin collecting data. You will be prompted to enter the volume reading from your syringe. Type this value into the box and press "Enter and Continue". The computer will plot the pressure (torr) and volume (mL) and then prompt you for the next volume reading. Repeat this process for 10 measurements adjusting the volume of the syringe in approximately 1 mL increments. Do not remove the plastic tubing while adjusting the volume of the syringe. When you have made 10 measurements, press the "Stop" button to stop data collection. Record the pressure values for each volume on your data sheet. For those measurements where you have two pressure readings for one volume, compute the average pressure.

Part B: Pressure vs. Temperature

In this part of the experiment, you will measure the change in pressure of a constant volume of gas as it is heated and cooled. The gas is contained in a closed syringe fitted with a black rubber stopper. Plastic tubing at the end of the syringe is connected to the pressure port on the MicroLab unit. Check that the black stopper is fitted securely. A temperature sensor is atttached to the outside of the plastic syringe so that the probe is pointing towards the black rubber stopper. Rubber bands are used to hold the temperature probe in place on the outside of the syringe. Figure 1 illustrates the apparatus for collecting temperature and pressure data.

The syringe/temperature probe assembly will be successively placed in water baths at least four different temperatures in order to determine the affect of temperature on gas pressure. The four water baths should contain: (1) ice water, (2) water at room temperature, and (3) hot tap water between 35°C and 50°C and (4) hot tap water between 60°C and 75°C . Use large 600 mL beakers for the baths. The water level should be sufficient to cover the syringe. Be careful that the plastic tubing is securely on the syringe to prevent water from entering the syringe.

From the menu click on the Gas Laws tab and select the Pressure vs. Temperature Relationship experiment template. Begin by immersing the syringe in the ice water bath so that the black stopper is facing down into the beaker of water. Allow the temperature to equilibrate for about two minutes. After two minutes, check all of your connections and press Start. You will be prompted to press "Continue" when T and P are stabilized. By clicking "Continue", the current values of temperature (°C) and pressure (torr) will be recorded and plotted. Move the test tube to the room temperature water bath and wait for the temperature to equilibrate and then press "Continue".

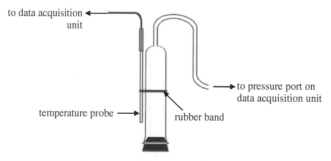

FIGURE 1—*Pressure and Temperature Apparatus*

Repeat the process with the other water baths. When the last measurement is made, press the "Stop" button. Transfer the temperature and pressure data you collected to your data sheet.

Data Analysis

In this part of the experiment, you will graph the data you obtained. Using data from Part A, create a graph of Pressure vs. Volume. Draw a curve through the data points. Now, create a graph of Pressure vs. 1/Volume. Draw a "best fit" straight line through these data points. Finally, using data from Part B, create a graph of Pressure vs. Temperature (K). Draw a "best fit" straight line through these data points.

Safety Precautions

There are no special safety precautions associated with this experiment.

Waste Management

There are no chemical wastes generated by this experiment.

Gas Laws Prelab Questions

Name ______________________________ **Section** ______________________________

1. A student connects a syringe to a pressure sensor and begins pushing and pulling on the plunger of the syringe while recording the pressure readings on a computer. A graph of this activity is shown below.

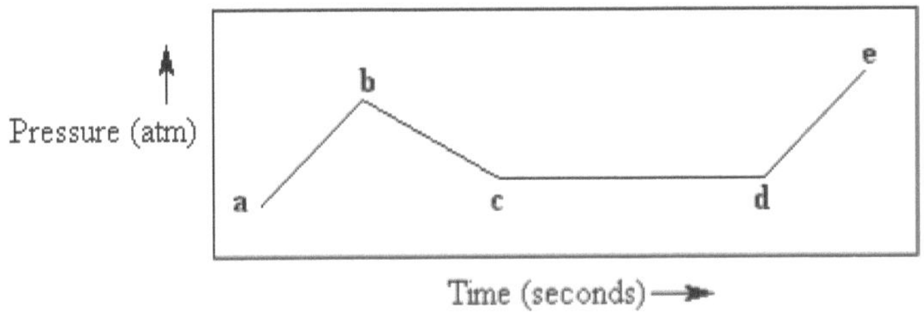

Draw an arrow so that the point of the arrow indicates the portion of the graph above that corresponds to the plunger being pulled out of the syringe (from a lesser to greater volume value).

2. A student connects a stoppered test tube to a pressure sensor and places the test tube in a series of four beakers containing water at different temperatures: 30 °C, 50 °C, 60 °C, and 80 °C. A computer continuously records the pressure in the test tube. A graph of this activity is shown below.

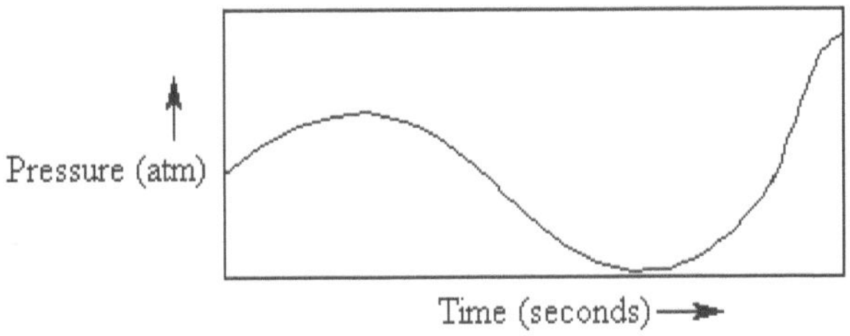

Arrange the four temperatures in the order that corresponds to the graph of pressure vs. time.

Gas Laws Data Sheet

Name _____ **Section** _____

Name of Lab Partner(s) _____

Part A: Boyle's Law: Pressure vs. Volume

Volume (mL)	1/Volume (mL^{-1})	Pressure (torr)

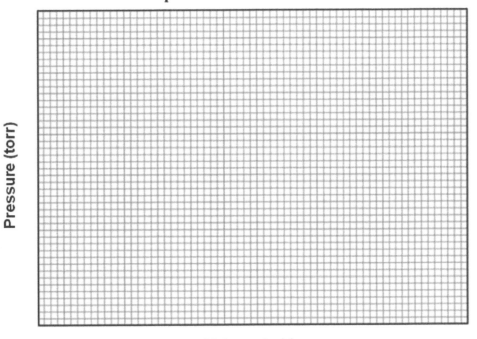

Graph 1: Pressure vs. Volume

Graph 2: Pressure vs. 1/Volume

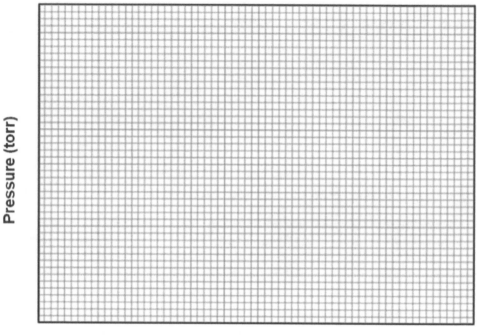

Pressure (torr)

1/Volume (mL⁻¹)

1. (a) Which graph (P vs. V or the graph of P vs. 1/V) would be best for predicting the pressure of the gas in the syringe when the volume is 2 mL? (b) Why?

2. According to your graph of P vs. 1/V, as the pressure of a gas goes up, the volume of the gas goes down. Explain why it is physically impossible, even under conditions of extremely high pressure, for the volume of a sample of a real gas to be reduced to zero.

Part B: Pressure vs. Temperature

Temperature (°C)	Temperature (K)	Pressure (torr)

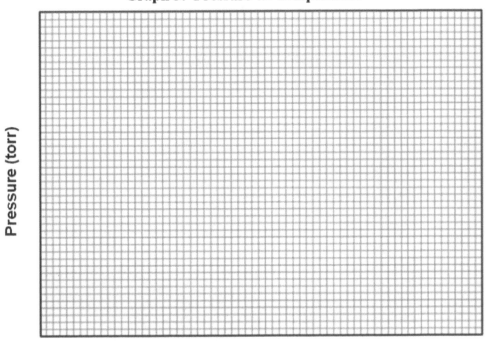

Graph 3: Pressure vs. Temperature

3. (a) Explain how you would modify your graph of P vs. T in order to determine the value of T when P = 0.
    (b) What is the significance of this temperature?

Gas Laws Review Questions

Name _____ **Section** _____

1. Consider the following graphs.

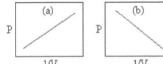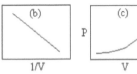

   Boyle's Law states the relationship between **pressure** and **volume** for a fixed amount of gas at constant temperature and is expressed as $P = k/V$ (k is a constant). Which graph best represents Boyle's Law? Why?

2. Consider the following graphs.

   The relationship between **volume** and **temperature** for a fixed amount of gas at constant pressure is known as **Charles' Law, $V = kT$** (k is a constant). Which graph do you think best represents Charles' Law? Explain how, using the equipment provided for this experiment, you would collect data for a Charles' Law experiment.

Gas Laws Supplemental Questions

Name _____ **Section** _____

Hess's Law

Introduction

There are a very large number of chemical reactions that have never been investigated. There may be good reasons for not performing certain chemical reactions. For example, some reactions may pose great risks to the environment or to living organisms. Other reactions may be too costly in terms of time or money. Studying the chemical reactivity of diamonds would certainly be an expensive undertaking. Fortunately, there are methods that allow us to study certain aspects of chemical reactions without ever having to perform the reactions.

Hess's Law (also known as the **Law of Constant Heat Summation**) is an example of a method that allows chemists to study the **thermodynamics** of a chemical reaction without having to perform the reaction itself. Germain Henri Hess discovered this fundamental law of nature in 1840 while studying the heat evolved when acids and bases are mixed with water. Hess's Law states that the **enthalpy change (ΔH)** of individual steps in a process can be added or subtracted to determine the overall enthalpy change of the process. Hess's Law is a direct consequence of the **First Law of Thermodynamics**, which requires that the enthalpy change of a reaction be the same regardless of the reaction pathway that leads from reactants to products. Thus, the net amount of heat liberated or absorbed in a chemical process is the same, regardless of whether the process is performed in one step or multiple steps.

In today's experiment, you will use Hess's Law to determine the enthalpy change (ΔH) of a chemical reaction that cannot be easily measured by experimental methods. This reaction is represented by equation (1) below.

$$CaCO_{3(s)} + H_2O_{(l)} \longrightarrow Ca(OH)_{2(s)} + CO_{2(g)} \qquad \Delta H_1 \qquad \text{[Eqn. 1]}$$

You will determine the value of ΔH for this reaction by measuring the enthalpy changes of two related chemical reactions. These reactions are represented by equations (2) and (3) below.

$$CaCO_{3(s)} + 2\,HCl_{(aq)} \longrightarrow CaCl_{2(aq)} + H_2O_{(l)} + CO_{2(g)} \qquad \Delta H_2 \qquad \text{[Eqn. 2]}$$

$$Ca(OH)_{2(s)} + 2\,HCl_{(aq)} \longrightarrow CaCl_{2(aq)} + 2\,H_2O_{(l)} \qquad \Delta H_3 \qquad \text{[Eqn. 3]}$$

Equation (1) can be obtained by adding equation (2) to the reverse of equation (3). If we reverse equation (3), we must also change the sign of the enthalpy change for the reaction. According to Hess's Law, if we add the equations for two reactions together, we can also add their enthalpy changes together. Thus, the enthalpy change for reaction (1) is simply the sum of the enthalpy change for reaction (2) and the reverse of reaction (3).

$$\Delta H_1 = \Delta H_2 + (-\,\Delta H_3)$$

To determine the enthalpy changes (ΔH_2 and ΔH_3) of reactions (2) and (3), you must measure the **heat (q)** released by each reaction. You will do this using the same techniques and concepts you used in the experiment on calorimetry.

(You might want to review the Introduction to *Calorimetry*.) Recall that the heat exchanged in a process can be calculated from the following equation

$$q = C_s \times g \times \Delta T$$

where **g** is the mass of the substance in grams, the **specific heat, C_s,** of the substance is expressed in units of J/g·°C, and ΔT is the temperature change the substance undergoes ($T_{final} - T_{initial}$).

The two reactions you will study in this experiment take place in aqueous solution. Each reaction is exothermic, so the heat released by the reaction is gained by the solution. This causes the temperature of the solution in the calorimeter to rise. By knowing the mass of solution in the calorimeter and the temperature change the solution undergoes, we can calculate the heat gained by the solution. (We are assuming that the specific heat of the solution in the calorimeter is essentially the same as the specific heat of water.)

$$q_{soln} = (4.18 \text{ J/g·°C}) (g_{soln}) (\Delta T_{soln})$$

The heat gained by the solution is equal in magnitude but opposite in sign to the heat lost by the reaction. Therefore, we can calculate $q_{reaction}$ from the following relationship.

$$q_{reaction} = - q_{soln}$$

Finally, the enthalpy change of each reaction, $\Delta H_{reaction}$, can be determined by dividing the heat lost by the reaction ($q_{reaction}$) by the number of moles of reactant used in the reaction (either $CaCO_3$ or $Ca(OH)_2$). For reaction (2), this calculation takes the following form.

$$\Delta H_{reaction(2)} = \frac{q_{reaction(2)}}{\text{moles } CaCO_3}$$

Procedure

Obtain two coffee cups with one lid to use as your coffee cup calorimeter. Place one coffee cup inside the other. Place the dry calorimeter cup on a balance and weigh it. Record this value on your data sheet. Add approximately 0.90 g of $CaCO_3$ to the calorimeter cup. Record the exact mass of your $CaCO_3$ on your data sheet.

After adding $CaCO_3$ to the calorimeter cup, add approximately 5.0 mL of DI water. Gently swirl the cup to mix the calcium carbonate and water. Place a temperature probe through the lid of the calorimeter into the water and $CaCO_3$ mixture. Continue to swirl the contents of the calorimeter and monitor the temperature until a stable reading is obtained. Record this temperature as $T_{initial}$ on your data sheet.

Obtain approximately 20.0 mL HCl. Lift the calorimeter lid and slowly pour the HCl into the calorimeter by allowing the hydrochloric acid to flow along the side of the cup. Continue monitoring the temperature while you swirl the cup to ensure complete reaction. You should observe that the temperature rises and then levels off. Record the temperature where it levels off as T_{final} on your data sheet. Remove the temperature probe and weigh the calorimeter again and record this value on your data sheet. This will be used to determine the total mass of solution in your calorimeter.

Take your calorimeter apart and thoroughly rinse the inner cup. Dry it completely. Place the dry calorimeter cup on a balance and weigh it. Record this value on your data sheet. Weigh approximately 0.60 g $Ca(OH)_2$ into the calorimeter cup. Record the exact mass of your $Ca(OH)_2$ on your data sheet.

After adding $Ca(OH)_2$ to the calorimeter cup, add approximately 5.0 mL of DI water. Gently swirl the cup to mix the calcium hydroxide and water. Place a temperature probe through the lid of the calorimeter into the water and $Ca(OH)_2$ mixture. Continue to swirl the contents of the calorimeter and monitor the temperature until a stable reading is obtained. Record this temperature as $T_{initial}$ on your data sheet.

Obtain approximately 20 mL of HCl and slowly add it to the mixture in the calorimeter. Gently swirl the calorimeter while monitoring the temperature. Record the temperature where it levels off as T_{final} on your data sheet. Remove the temperature probe and weigh the calorimeter again and record this value on your data sheet. This will be used to determine the total mass of solution in your calorimeter.

Calculate ΔT, $q_{reaction}$, and $\Delta H_{reaction}$ for reactions (2) and (3). Enthalpy values are expressed in units of kJ/mol, so you will need to convert from joules to kilojoules when calculating the enthalpy changes for reactions (2) and (3). Use the values you calculate for ΔH_2 and ΔH_3 to calculate ΔH_1.

Safety Precautions

The concentration of HCl used in this experiment will eat holes in your clothing and will cause burns to your skin. Exercise caution while measuring and mixing HCl.

Waste Management

The products of the reactions in this experiment can be poured down the sink and flushed with water. Be sure to wash the cup thoroughly so that nothing remains on the sides. Rinse the calorimeter with DI water and return it to the reagent bench.

Hess's Law Prelab Questions

Name _____ **Section** _____

1. A thermometer placed in a solution undergoing a chemical reaction indicates an increase in temperature as the reaction proceeds. Is this _reaction_ endothermic or exothermic? Describe if heat energy is lost or gained from the reaction (the system) to the surroundings. What is the sign of the enthalpy change (ΔH) of this reaction?

2. A student performs a reaction and determines the enthalpy change (ΔH) to be 31.4 kJ. Will the temperature of the _surrounding solution_ increase or decrease as a result of this chemical process?

3. If you hold 3 grams of ice in your hand at room temperature, your hand will become cold. a) Is the reaction $H_2O(s) \longrightarrow H_2O(l)$ endothermic or exothermic? b) In which direction does heat flow?

Hess's Law Data Sheet

Name _____ **Section** _____

Name of Lab Partner(s) _____

Reaction (2): $CaCO_{3(s)} + 2HCl_{(aq)} \longrightarrow CaCl_{2(aq)} + H_2O_{(l)} + CO_{2(g)}$

a) Mass of empty calorimeter _____

b) Mass of calorimeter with $CaCO_3$, water, and HCl _____

Mass $CaCO_3$	Moles $CaCO_3$	$T_{initial}$	T_{final}	ΔT (°C)	g (mass of soln)	$q_{solution}$ (joules)	$q_{reaction(2)}$ (joules)	ΔH_2 (kJ/mol)

Reaction (3): $Ca(OH)_{2(s)} + 2HCl_{(aq)} \longrightarrow CaCl_{2(aq)} + 2H_2O_{(l)}$

c) Mass of empty calorimeter _____

d) Mass of calorimeter with $Ca(OH)_2$, water, and HCl _____

Mass $Ca(OH)_2$	Moles $Ca(OH)_2$	$T_{initial}$	T_{final}	ΔT (°C)	g (mass of soln)	$q_{solution}$ (joules)	$q_{reaction(3)}$ (joules)	ΔH_3 (kJ/mol)

1. Combine the above reactions (2) and (3) to give the reaction of interest (1) in the space below.

2. Appropriately combine the measured enthalpies of reactions (2) and (3) to predict ΔH_1 for the following reaction:

Reaction (1): $CaCO_{3\,(s)} + H_2O_{(l)} \longrightarrow Ca(OH)_{2\,(s)} + CO_{2\,(g)}$ $\Delta H_1 = $ _____

3. Based on the predicted enthalpy, indicate whether **Reaction (1)** is endothermic or exothermic (circle the type).

Hess's Law Review Questions

1. Write a balanced equation for the formation of $CO_{2(g)}$ from $C_{(s)}$ and $O_{2(g)}$. Calculate the enthalpy change for this reaction using the following data (at 25 °C):

    $$C_{(s)} + \tfrac{1}{2} O_{2(g)} \longrightarrow CO_{(g)} \qquad\qquad \Delta H = -111 \text{ kJ}$$
    $$CO_{(g)} + \tfrac{1}{2} O_{2(g)} \longrightarrow CO_{2(g)} \qquad\qquad \Delta H = -394 \text{ kJ}$$

    Is this reaction endothermic or exothermic?

2. a) Explain how ΔT would be affected if a greater amount of surrounding solvent (water) is used, assuming the mass of salt remains constant?

    b) Explain how $q_{reaction}$ would be affected if a greater amount of surrounding solvent (water) is used? Explain.

3. If the following enthalpies are known:

    $$A + 2B \longrightarrow 2C + D \qquad\qquad \Delta H = -95 \text{kJ}$$
    $$B + X \longrightarrow C \qquad\qquad \Delta H = +50 \text{kJ}$$

    What is ΔH for the following reaction?

    $$A \longrightarrow 2X + D$$

Hess's Law Supplemental Questions

Name _____ **Section** _____

Intermolecular Forces: Evaporation of Alcohols

Introduction

Intermolecular forces (abbreviated IMF) are attractive forces that exist *between* molecules and influence the physical properties of pure substances and mixtures. A few of the physical properties that are influenced by intermolecular forces include melting point, boiling point, viscosity, surface tension, and vapor pressure. The rate at which a liquid evaporates is a function of its vapor pressure, which in turn is a function of its intermolecular forces. Liquids with high vapor pressures have weak intermolecular forces holding the molecules together in the liquid. Consequently, molecules with sufficiently high kinetic energy can break away from the liquid and enter the gas phase. This lowers the average kinetic energy of the remaining molecules in the liquid causing the temperature of the liquid to decrease.

The decrease in temperature of a liquid as it evaporates is called **evaporative cooling**. The rate at which an evaporating liquid cools can be used to estimate the strength of the intermolecular forces in the liquid. In today's experiment, an electronic temperature sensor will be placed in various liquids. When the sensor is removed from a liquid, the liquid adhering to the sensor begins to evaporate causing the temperature of the sensor to decrease. The size of this decrease is related to the strength of the intermolecular forces in the liquid.

Two types of organic compounds will be used in this experiment: **alkanes** and **alcohols**. Alkanes are hydrocarbons (organic compounds containing only carbon and hydrogen) and include pentane, C_5H_{12}, and hexane, C_6H_{14}. Alcohols contain the –OH functional group and include methanol, ethanol, propanol, and butanol.

By examining the molecular structures of the alkanes and alcohols used in this experiment, you can determine the types of intermolecular forces present in each substance. The two most important intermolecular forces in this experiment are **hydrogen bonding** and **dispersion forces**. Hydrogen bonding occurs among molecules that contain a hydrogen atom covalently bonded to N, O or F. Dispersion forces exist in all substances and increase as the molecular weight of a substance increases.

Procedure

Part A:

Temperature data for this experiment will be collected using a computer, the MicroLab data acquisition unit, and temperature probe. Cut a 1" square piece of filter paper and roll it into the shape of a cylinder. Wrap the paper cylinder around the tip of the temperature probe and secure it with a small rubber band. The end of the paper should be even with the end of the probe. (Hint: first slide the rubber band onto the probe, wrap the paper around the probe, and then slide the rubber band over the wrapped paper).

Pour 2–5 mL of **ethanol** into a medium test tube and then place the temperature probe in the ethanol. Make sure there is enough liquid in the test tube to completely cover the paper wrapped around the probe. Secure the test tube in a test tube rack or beaker so that it does not tip over.

Open the "Intermolecular Forces-Evaporative Cooling CHML 102" experiment under the "Time and Temperature" prompt in the MicroLab program. Before selecting the "Start" prompt, allow the temperature probe to remain in the liquid for at least 45 seconds. Without removing the temperature probe from the liquid, begin collecting temperature data by clicking the "Start" prompt on the computer. Collect temperature data for 45 seconds with the probe in the ethanol, then remove the probe from the ethanol and lay it on the laboratory bench so that the probe tip extends 5 cm over the edge of the lab bench (this prevents any heat exchange between the lab bench and the probe). The temperature will decrease due to the evaporation of the liquid on the filter paper. Continue to collect temperature data until a minimum temperature has been reached, then click the "Stop" prompt so that the data may be reviewed.

Examine the data on the computer screen and calculate the difference between the maximum and minimum temperatures reached during the experiment (ΔT) as shown in Figure 1. Record this value on your data sheet. Slide the rubber band up the temperature probe and dispose of the filter paper in the trashcan.

Repeat the procedure using **1-propanol** in place of ethanol. Use a new piece of filter paper whenever you change liquids. Recycle the rubber bands. Repeat the procedure, this time using **n-pentane**.

Begin to think about any correlation you observe between values of ΔT and the types of intermolecular forces present in ethanol, 1-propanol, and n-pentane. Keep in mind that a large ΔT indicates a high rate of evaporation and weak intermolecular forces.

Part B:

Based on the ΔT values you obtained in Part A and information from the table you completed in the Prelab Questions, predict the size of ΔT for **methanol** and **1-butanol**. Compare hydrogen-bonding capability, molecular weight, and other intermolecular forces to those in ethanol, 1-propanol, and n-pentane. Record your predicted ΔT as a value less than (<) or greater than (>) the ΔT values for the liquids in Part A. It is not important that you predict an exact ΔT value, simply state whether your predicted value is greater than, less than, or equal to the previous ΔT values.

Now confirm or disprove your predictions for methanol and 1-butanol by experimentally determining ΔT for these two alcohols using the procedure described in Part A. Answer the questions that follow Part B.

Part C:

Based on the ΔT values you measured for all five liquids in Parts A and B, predict the ΔT value for **n-hexane**. Record your predicted value of ΔT. Test your prediction by experimentally determining ΔT for n-hexane. Complete the questions that follow Part C.

Safety Precautions

Be careful not to get any of the liquids used in this experiment on the computer. Avoid inhaling the vapors of the liquids used in this experiment. Avoid skin contact with methanol, which is toxic and readily absorbed through the skin. Gloves are available for use if you prefer to use them. Wash your hands thoroughly after the experiment.

Waste Management

Place all chemicals in the container in the fume hood labeled "Organic waste". Return the small rubber bands to the beaker located on the front reagent shelf. Filter papers may be thrown in the solid waste receptacle.

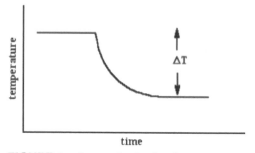

FIGURE 1—*Evaporative Cooling Curve*

Intermolecular Forces Prelab Questions

Name _____ **Section** _____

Complete the following table, which lists the substance used in the experiment:

Substance	Formula	Structural Formula	Molecular Weight	Hydrogen bonds? (yes or no)
methanol	CH_3OH			
ethanol	C_2H_5OH			
1-propanol	C_3H_7OH			
1-butanol	C_4H_9OH			
n-pentane	C_5H_{12}			
n-hexane	C_6H_{14}			

Intermolecular Forces Data Sheet

Name _____ **Section** _____

Name of Lab Partner(s) _____

Part A

Substance	$T_{maximum}$	$T_{minimum}$	ΔT
ethanol			
1-propanol			
n-pentane			

Part B

Substance	$\Delta T_{predicted}$	$T_{maximum}$	$T_{minimum}$	$\Delta T_{experiment}$
methanol				
1-butanol				

a) Justify your predictions of ΔT for methanol and 1-butanol.

b) Did your experimental results confirm your predictions? In your explanation, describe the types of IMF of the compounds you are comparing.

Part C

Substance	$\Delta T_{predicted}$	$T_{maximum}$	$T_{minimum}$	$\Delta T_{experiment}$
n-hexane				

a) Justify your prediction of ΔT for n-hexane.

b) Did your experimental results confirm your prediction? In your explanation, describe the types of IMF of the compounds you are comparing.

Intermolecular Forces
Review Questions

Name _____ **Section** _____

1. Two of the liquids you used, n-pentane and 1-butanol, have nearly the same molecular weights, but significantly different ΔT values. Using data collected from your experiment, explain the difference in the ΔT values of these substances based on their intermolecular forces.

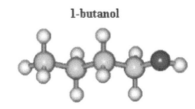

n-pentane   1-butanol

2. Glycerine (glycerol) $CH_2-OHCH-OHCH_2-OH$ has a greater resistance to flow (viscosity = 94) than ethanol CH_3CH_2OH (viscosity = 1.08). Which of these two liquids will have the largest temperature loss due to evaporative cooling and why?

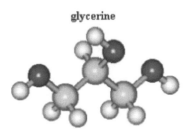

glycerine

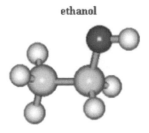

ethanol

Intermolecular Forces
Supplemental Questions

Name _____ **Section** _____

Kinetics

Introduction

Chemical **kinetics** is the branch of chemistry concerned with the speed (or rate) at which chemical reactions occur. Some chemical reactions proceed at such slow rates that millions of years would be required to notice any appreciable change in reactant concentrations. Other reactions proceed so rapidly that their rate is impossible to measure with current methods. In addition to measuring rates of chemical reactions, chemical kinetics is also concerned with studying the factors that contribute to increasing or decreasing the rates of chemical reactions.

In this experiment and the one that follows, you will study two factors that affect the rates of chemical reactions: (a) reactant concentrations and (b) temperature. Virtually every chemical reaction will proceed more rapidly if the concentrations of reacting species are increased. (Only zero order reactions are not affected by changes in reactant concentrations.) Consider the following generalized chemical reaction:

$$aA + bB + cC \longrightarrow dD + eE$$

The rate of this reaction can be expressed as a change in concentration of any one of these species with time. By convention, the rate of a reaction is expressed as a change in concentration of one of the reactants with time. Furthermore, the rate of a reaction is always positive, with units of **mol/L·sec**. For the reaction shown above, the rate could be expressed as

$$Rate = -\frac{\Delta[A]}{\Delta t}$$

The negative sign is required to keep the rate a positive number since the concentration of A is decreasing during the reaction. The rate of the reaction is related to the concentration of reactants by an expression known as the **rate law**. For the reaction shown above, the rate law would be

$$Rate = -\frac{\Delta[A]}{\Delta t} = k[A]^x[B]^y[C]^z$$

The letter **k** refers to the **rate constant** for the reaction. The letters x, y, and z are the **orders** of the reaction with respect to reactants A, B, and C. These are generally small whole numbers like 0, 1, 2, or 3. The **overall order** of the reaction is the sum of x, y, and z. For example, assume x = 1, y = 2, and z = 2. The reaction would be first order with respect to reactant A, and second order with respect to reactants B and C. The reaction would be fifth order overall. The order of a reaction indicates how the rate of the reaction changes with changes in reactant concentrations. For the example just given, assume the concentrations of reactants A and C are held constant while the concentration of reactant B is doubled. Because the rate of the reaction varies with the square of the concentration of B, the rate of reaction would increase four-fold.

In this experiment, you will measure the rate of reaction (1) with various concentrations of BrO_3^-, I^-, and H^+. The complete reaction is

$$6\ I^- + BrO_3^- + 6\ H^+ \longrightarrow 3\ I_2 + Br^- + 3\ H_2O \qquad \text{[Eqn. 1]}$$

The rate data will be used to determine the order of the reaction with respect to each reactant. Once the orders (values for x, y, z in the rate law expression (2) below) are known, the rate constant (k) for the reaction can be calculated. The rate law for this reaction is given by

$$\text{Rate} = k\ [\mathbf{BrO_3^-}]^x\ [\mathbf{I^-}]^y\ [\mathbf{H^+}]^z \qquad \text{[Eqn. 2]}$$

The rate of this reaction at room temperature is slow enough to be easily measured. However, the problem is to find some way of determining how much BrO_3^- is used up in a certain length of time. This problem is solved by introducing into the reaction flask a second reaction called a **clock reaction**. A clock reaction is one that changes color when a predetermined amount of reagent has been consumed. The clock reaction used in this experiment is shown in equation (3).

$$I_2 + 2\ S_2O_3^{2-} \longrightarrow 2\ I^- + S_4O_6^{2-} \qquad \text{[Eqn. 3]}$$

The reaction between iodine (I_2) and thiosulfate ($S_2O_3^{2-}$) is nearly instantaneous. As I_2 is produced in reaction (1), it is immediately consumed in reaction (3). When all of the $S_2O_3^{2-}$ is gone, the concentration of I_2 starts to increase in the solution. The addition of starch to the solution produces a deep blue color with the I_2. Therefore, the appearance of a blue color in the reaction flask indicates when a known amount of BrO_3^- has reacted.

Procedure

Part A: Determination of Reaction Orders

A total of five trials will be conducted. Each trial will require adding the contents of Flask B to the contents of Flask A and measuring the time it takes for the blue color to appear. The contents of each flask for each trial are shown in Table 1. The ions outlined in bold: I^-, BrO_3^-, and H^+ are the reactants shown in equation (1) which are being varied in order to determine how changes in their concentration affects the rate of the reaction. Note the concentration of thiosulfate ion ($S_2O_3^{2-}$) remains constant throughout the five trials. Water is added to bring the total volume in each trial up to a volume of 25 mL and does not affect the rate because it is not a reactant.

Each trial is run in exactly the same manner, therefore only the procedure for Trial 1 is described. Using your 10 mL graduated cylinder, measure the quantities described in Table 1 into Flask A and Flask B. Be sure to rinse the graduated cylinder with deionized water before the addition of each reagent. Dry the cylinder to avoid diluting

Table 1 Reagent Volumes (mL)

	Flask A (250mL)			Flask B (125mL)	
Molarity:	0.010M KI	0.0010M Na₂S₂O₃	H₂O	0.040M KBrO₃	0.10M HCl
Trial 1	5	5	5	5	5
Trial 2	5	5	0	10	5
Trial 3	10	5	0	5	5
Trial 4	5	5	0	5	10
Trial 5	2.5	5	7.5	2.5	7.5

the concentrations of the reagents. **Place 4 drops of starch into Flask B.** *All trials must be conducted at the same temperature.* Record this temperature on the data sheet.

When you are ready to begin, note the position of the sweep second hand on the wall clock and add the contents of Flask B to Flask A. *Swirl the flask continuously and record the time when it turns blue.* If the blue color does not appear in 5 minutes, rinse the flask and start over. Repeat this process for Trials 2, 3, and 4. Do not do Trial 5 at this time. Be sure to rinse the flasks with deionized water before reusing them.

The calculations for Part A require you to first determine the concentrations of BrO_3^-, I^-, and H^+ present in the reaction flask in each of the trials since the initial concentrations shown in Table 1 are diluted when the reagents are mixed. This is done by multiplying the concentration of each reagent by the volume used (in mL) and dividing by 25 (the total volume of the solution).

$$M_{final} = \frac{(mL \times M_{initial})}{25\ mL}$$

These calculations have been performed for you for Trial 1. The rate of the reaction in each trial is found by dividing the change in concentration of BrO_3^- by the time it took for this change to occur. The change in concentration of BrO_3^- is the *same* in each trial since the amount of $S_2O_3^{2-}$ is the *same* in each trial. (There is 1/6 mole of BrO_3^- reacted for every mole of $S_2O_3^-$ reacted.) This means that 3.3×10^{-5} mol/L of BrO_3^- reacted in each trial.

$$\text{Rate of reaction} = \frac{3.3 \times 10^{-5}\ \text{mol/L}}{\text{seconds required to turn blue}}$$

The order of the reaction with respect to each reactant is found by analyzing the data in the table in a manner similar to Prelab Question #2. The orders for this reaction are small whole numbers. After calculating the concentrations of the reactants and determining times for the first four trials, take your data sheet to your instructor before proceeding.

Part B: Determination of Rate Law

When you have determined the order of the reaction with respect to each reactant, write the expression for the rate law for this reaction.

Part C: Determination of Rate Constant

Using the rate law you determined in Part B, calculate the reaction rate constant for each trial. Determine the average rate constant over all four trials. Once you know the rate law, you can solve for the rate constant for a given trial using the concentrations listed in Table 1. Substitute the concentrations and experimentally determined rate for that trial into the rate law you determined in step B above. Raise the concentrations of reactants to the appropriate power based on the order for that reactant and solve for the rate constant.

Part D: Estimating Time for Trial 5

Using your rate law, the average rate constant k, and the concentrations of reactants described in Trial 5, calculate the time it should take for Trial 5 to turn blue. Once you have done this, mix the solutions for Trial 5 and measure the time it actually takes. Compare this to your calculated value.

Safety Precautions

Use caution when working with HCl as it is may cause burns.

Waste Management

Flush all wastes from this experiment into the sink with plenty of tap water.

Kinetics Prelab Questions

Name _____ **Section** _____

1. a) Write the Rate Law for the following reaction given that the order of A = 3, B = 1, and C = 0

$$A + B + C \longrightarrow D + E$$

b) If the concentration of C is doubled, what will happen to the rate of the reaction?

2. The table below contains concentration and rate data for a reaction involving three reactants. Use these data to determine the overall order of the reaction and the order with respect to M, N and P.

Trial	[M]	[N]	[P]	Rate
1	0.01	0.01	0.02	2.6
2	0.02	0.01	0.02	5.2
3	0.02	0.02	0.02	5.2
4	0.01	0.01	0.01	0.32

Order with respect to **M** _____

Order with respect to **N** _____

Order with respect to **P** _____

Overall order _____

Kinetics Data Sheet

Name _____ **Section** _____

Name of Lab Partner(s) _____

Temperature _____

Part A

Trial	$[BrO_3^-]$	$[I^-]$	$[H^+]$	Time (sec)	Rate (mol/L·s) $[3.3 \times 10^{-5}]/t$
1	0.0080	0.0020	0.020		
2					
3					
4					
5					

Order with respect to BrO_3^- _____

Order with respect to I^- _____

Order with respect to H^+ _____ Instructor's initials _____

Part B

Using the determined orders, write the expression for the rate law in the space below:

Part C

Determine the value of k for each trial.

k_1 _____ k_2 _____ k_3 _____ k_4 _____

What is the average value of k for all four trials?

$k_{average}$ _____

Part D

Calculate the time required for Trial 5 to turn blue.

Calculated time _____

What was the actual time measured for Trial 5? Measured time _____

Kinetics Review Questions

Name _____ Section _____

1. The following data were collected at 20 °C for the reaction below.

$$\text{Acetone} + H^+ + Br_2 \longrightarrow \text{bromoacetone}$$

The time to produce a small but constant amount of bromoacetone was measured while varying the concentrations of Br_2, acetone, and H^+.

Br_2 (M)	acetone (M)	H^+ (M)	Rate
0.010	0.25	0.10	35
0.010	0.13	0.10	18
0.020	0.25	0.10	36
0.020	0.13	0.40	71

What is the order of the reaction with respect to Br_2, acetone, and H^+?

2. For the reaction A + B ⟶ C, which one of the following represents the correct graph of the concentration of A with time? (Assume this reaction is a first-order reaction.)

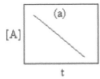

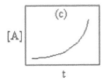

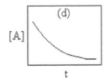

Kinetics Supplemental Questions

Name _____ Section _____

Molecular Shapes

Introduction

One of the most important advances that occurred in chemistry during the last 100 years was the development of theoretical models that predict the shapes of covalently bonded molecules. A knowledge of molecular structure not only allows us to explain why known substances behave the way they do, but it also allows us to predict the physical and chemical properties of substances not yet made.

Most current models are based on the work of G.N. Lewis who, in 1916, proposed that bonds in molecules consisted of pairs of electrons. These **valence electrons** that are involved in the chemical bonding are located in the incomplete outer shell of the atom. We now label these outermost valence electrons as *Lewis electron-dot symbols*, or as *Lewis symbols*. The Lewis symbol for an element uses a chemical symbol for the element in addition to a dot to designate each valence electron that element has in its outermost shell. The following example of the Lewis dot symbol of nitrogen shows the use of the chemical symbol of nitrogen, N, and the use of five dots to signify the five outermost valence electrons, as in the electron configuration $[He]2s^2 2p^3$:

The symbol of the element is written so that each side can accommodate 2 electrons, or an electron pair. To determine the number of valence electrons, one can look at the group number of that element in the periodic table. To follow this rule, we note that the Cl atom is in the 7A column on the periodic table, which indicates that this atom (and those in the same column) will have 7 outer valence electrons.

Lewis also suggested that most atoms in molecules tend to have eight electrons in their outermost valence shell (except hydrogen which can only have two valence electrons). We know of many molecules that contain atoms that do not have eight electrons in their valence shell, but so many molecules do conform to this simple concept that it has become known as the **octet rule**. The octet rule states that atoms tend to gain, lose, or share electrons until they are surrounded by eight valence electrons, which is the same number of electrons as the noble gas nearest to that element in the periodic table. Thus, a Cl atom must gain only one electron to gain a full octet and achieve the noble gas configuration of Ar.

Covalently bonded atoms are able to achieve their octet by sharing their electrons among the atoms in the molecule. *Lewis dot structures* are a simple representation of a molecule that shows the sharing of electrons between the atoms in that molecule (in which a single chemical bond is produced by sharing two electrons). For example, the

Lewis dot structure of the Cl_2 molecule is illustrated below where the chemical bond between them is a shared electron pair (note that each Cl atom has a full octet):

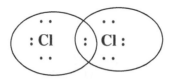

Covalent molecules sharing an electron pair form a single bond. However, many molecules can form double or triple bonds to complete their octets. In double bonds two pairs of electrons are shared; likewise, triple bonds are composed of three pairs of electrons shared between two atoms.

The following **rules** will help you construct Lewis dot structures:

1. Determine the number of valence electrons for each atom in the molecule or the ion.

2. Total the number of valence electrons in the molecule or ion (add electrons for negative ions, and subtract electrons for positive ions).

3. Identify the center atom and surrounding atoms. Write the chemical symbol for each atom in the molecule.

4. Divide the total number of valence electrons by 2, the number of electrons in each bond, to give the number of bonds in the molecule or ion. Draw a single bond between each atom and the center atom.

5. Determine the possible structure connecting the atoms in the molecule or ion and distribute the unshared pairs of electrons around each atom so as to obey the octet rule. Signify each pair of unshared electrons by drawing 2 dots.

    a) Keep in mind the *exceptions to the octet rule:* hydrogen can only share 2 electrons, beryllium can only share 4 electrons, and boron can only share 6 electrons.

    b) Some elements, such as P, S, and Xe, can have "**expanded octets**" which means they can accommodate more than 8 electrons.

6. Place any leftover electrons on the central atom even if this exceeds the octet of the center atom.

7. If the octet rule cannot be satisfied try to use multiple bonds.

Helpful Hints

A. H atoms should be attached to any O atom present or if no O is present, it should be attached to the central atom.

B. Terminal halogen atoms (F, Cl, Br, I) always have 4 electron pairs and form only single bonds.

C. Carbon atoms tend to form 4 bonds

D. Only C, N, O, and S ordinarily form double bonds

E. Nonbonding electron pairs on C atoms are very rare. C is almost never found at the end of a chain.

F. In molecules like XO_n, all O atoms are bonded to the X atom.

Two chemists working in England, Sidgwick and Nyholm, used Lewis' electron pair idea to develop the "**valence shell electron pair repulsion**" model (VSEPR; **Appendix B**) for predicting the geometry of covalently bonded molecules. This model held that the electron pairs in a molecule (both bonding and nonbonding) would try to get as far apart as possible. This attempt to minimize electron repulsion leads to the observed shape of the molecule. The most stable arrangements for two through six electron pairs around a central atom are shown in Table 1.

Table 1 *Electron Pair Distribution around a Central Atom*

Number of Electron Pairs	Dihedral Angle	Electron-Domain Geometry
2	180°	linear
3	120°	trigonal planar
4	109.5°	tetrahedral
5	120° and 90°	trigonal bipyramidal
6	90°	octahedral

It should be noted that multiple pairs of electrons shared between two atoms are counted as a single pair when determining molecular shapes. For example, consider the ammonia molecule (NH_3). The Lewis formula for ammonia is:

$$\cdot\,\cdot$$
$$H \;:\; N \;:\; H$$
$$H$$

Because there are four electron pairs around the central atom, the electron pairs will be arranged in the shape of a tetrahedron as shown below. The geometry of the compound if all the electron domains surrounding the central atom are considered is called the ***electron-domain geometry***.

In designating the molecular geometry, called the ***molecular geometry***, we consider only the actual atoms surrounding the central atom, and not the non-bonding electron pairs. For example, the molecular geometry of an ammonia molecule is trigonal pyramidal (a pyramid with a triangular base).

To determine the molecular and electron-domain geometry a summary is given in the following rules:

1. Count the number of valence electrons and draw the Lewis structure.

2. From the Lewis structure count the number of unshared electron pairs around the central atom and the number of electron pairs bound to the central atom. Multiple bonds are counted as a one bond to determine geometry.

3. Determine the arrangement of bonds and unshared electron pairs that will minimize electron pair repulsions. This shape will yield the electron-domain geometry.

4. Determine the shape of the molecule in terms of the position of the atoms including only the bonded pairs of electrons. This shape yields the molecular geometry.

5. Refer to "Appendix B: Summary VSEPR and Hybridization Table" for detailed information on hybridization, electron-domain and molecular geometry and pictures of molecular models.

One important piece of information that can be gained by knowledge of the structure of a molecule is whether or not the molecule is **polar**. A molecule is polar if there is unequal sharing of electron density within the molecule. This uneven distribution of negative charge leads to the development of partially positive ($\delta+$) and partially negative ($\delta-$) ends on the molecule. If the electron distribution in a molecule is symmetrical, then the molecule is **nonpolar**.

The electron density in a molecule can become unevenly distributed due to the attraction that highly electronegative atoms have for the electrons in the bonds they form. Examples of a polar and nonpolar molecule are shown below.

The water molecule is polar because opposite ends of the molecule have opposite charges, and there is an uneven sharing of electrons within that molecule. Consider the strong electron pull of the highly electronegative oxygen atom compared with the weak hydrogen atoms. The nonpolar carbon dioxide molecule exhibits an even distribution of electron density (the very electronegative oxygen atoms are pulling electron density equivalently and have equivalent partial negative charges).

To indicate the direction of the strongest electron pull, or the partially negative part of the molecule, an arrow called the **dipole moment** (+——►) is drawn so that the point is directed towards the negative side of the molecule, and the "cross" at the end of the arrow represents the positive side of the molecule. An example of a dipole moment used to demonstrate the uneven sharing of electrons is shown below. Note that dipole moments are only used on polar molecules.

$$\delta+ \qquad \delta-$$
$$H \text{ +} \longrightarrow Cl$$

If we consider the covalent bond formed in the HCl molecule above, we realize that the bond is a result of the overlap between a 1s orbital in hydrogen, and one of the 3p orbitals of the Cl. The overlapping orbitals in a bond in a simple structure such as HCl are easy to visualize. However, what about the overlapping of orbitals in a polyatomic structure such as $BeCl_2$? There are two bonds in this compound, and it is hard to discern which of the $2s^2$ electrons of beryllium overlaps with which p orbital of the Cl atoms. To create the two bonds in the $BeCl_2$ compound, the s and p orbitals actually mix to form a new type of orbital called "sp" **hybridized orbital**. Creating a new hybridized orbital allows the formation of a favorable overlap using a mixture of two or more types of atomic orbitals using a procedure called hybridization. The $BeCl_2$ molecule uses the hybridized "sp" orbital to create each bond between beryllium and chlorine atoms. A compound with 3 bonds uses "sp^2" hybridized orbitals for each bond; a compound with 4 bonds uses "sp^3" hybridized orbitals; a compound with 5 bonds uses "sp^3d" hybridization; and a compound with 6 bonds uses "sp^3d^2" hybridization.

In this experiment you will gain practice determining the geometries of various molecules and ions, determine the hybridization of the molecule or ion, and determine whether each exhibit polar characteristics.

Procedure

Part A

Draw the Lewis dot structures for the following compounds in the space given on the Data Sheet. Determine the number of bonding pairs, the number of lone pairs on the central atom, and the hybridization of the central atom.

Part B

Consider the list of molecular formulas provided in the following table. For each, draw a valid Lewis structure and predict the electron-domain and molecular geometries, the hybridization of the central atom, and whether the molecule is polar or non-polar. Finally, if the molecule is polar, draw an arrow on its structure indicating the direction of the dipole moment. As an example, the first one, oxygen difluoride, is completed for you.

Part C

Once you have completed the chart above in Part B, it is time to check your answers. Using a computer, go to http://cheminfo.chem.ou.edu/~mra/jmol/jmol.php. From the drop-down menu at the top, select the molecular formula for the compound you wish to check. The system will render that molecule in 3D (does it match the shape that you predicted?) and by checking the "molecular dipole" tab you can check whether the molecule is polar or nonpolar. In the space provided, correct any entries that you and your partner made and briefly explain why your initial answer was incorrect.

Molecular Shapes Prelab Questions

Name _____ **Section** _____

1. In the space below, draw the Lewis dot structure for the HI molecule. Give the shape of the molecule in electron-domain geometry and molecular geometry.

2. In the space below, draw the Lewis dot structure for the $BeCl_2$ molecule. Give the shape of the molecule in electron-domain geometry and molecular geometry.

3. Which of the following molecules contain polar bonds?

    a) H_2

    b) $SnCl_2$

    c) HBr

    d) CO_2

    e) BF_3

4. Circle the species below that violate the octet rule.

    a) PCl_5

    b) OH^-

    c) BeI_2

5. Draw the Lewis dot structure for the $ClBr_5$. What is the molecular geometry?

    a) trigonal bipyramidal

    b) square pyramidal

    c) square planar

    d) seesaw

Molecular Shapes Data Sheet

Name _____ **Section** _____

Name of Lab Partner(s) _____

Part A

Molecular Formula	Lewis Dot Structure	Number of Bonding Pairs	Number of Lone Pairs on Central Atom	Hybridization of Central Atom
CH_4				
NH_4^+				
H_3O^+				
CS_2				
$COCl_2$				
XeF_2				

Molecular Formula	Lewis Dot Structure	Number of Bonding Pairs	Number of Lone Pairs on Central Atom	Hybridization of Central Atom
N_2				
SO_2				
HCN				
PCl_5				
SF_4				
SO_4^{2-}				

Part B and Part C

Name	Molecular Formula	Lewis Structure	Electron-Domain Geometry	Molecular Geometry	Hybridization of Central Atom	Polar or Nonpolar?
Oxygen difluoride	OF_2		Tetrahedral	Bent	sp^3	Polar
Carbon dioxide	CO_2					
Xenon tetrafluoride	XeF_4					
Chlorine trifluoride	ClF_3					
Sulfur hexaflouride	SF_6					
Ammonia	NH_3					
Dichloro-methane	CH_2Cl_2					
Iodine pentabromide	IF_5					

Molecular Shapes Review Questions

Name _____ **Section** _____

1. Draw the Lewis dot structure and predict the molecular and electron-domain geometries of the following ions.

   a) NO_2^-

   b) CO_3^{2-}

   c) ClO_2^-

2. It is possible to write four resonance structures for the azide ion, N_3^-, all of which satisfy the octet rule (write them all). However, experimental evidence shows that the azide ion has a linear shape, with equal nitrogen-nitrogen distances. Which Lewis structure of N_3^- best represents this observation?

Molecular Shapes
Supplemental Questions

Name _____ **Section** _____

Molecular Weight Determination of Butane Gas

Introduction

The **Ideal Gas Law** expresses the relationship between the number of moles of a gas (**n**), the pressure of the gas (**P**), volume (**V**), and temperature (**T**) of the gas. The equation for this relationship is

$$PV = nRT$$

R, the ideal gas constant, has a value of 0.0821 if the pressure of the gas is given in atmospheres, the volume in liters, and the temperature in Kelvin degrees.

An important use of the Ideal Gas Law is in determining the molecular weights (*M*) of gases. The number of moles of gas, **n**, can be written in terms of the mass (**g**) of a gas sample and the molecular weight of the gas.

$$n = \frac{g}{M}$$

Substituting this expression for **n** in the Ideal Gas Law results in the following equation.

$$PV = \frac{gRT}{M}$$

Rearranging this equation to solve for molecular weight leads to the following relationship.

$$M = \frac{gRT}{PV}$$

It is relatively easy to measure the volume, temperature, and pressure of a gas, but not the mass of a gas. In this experiment, we overcome this difficulty by starting with a liquid, which is easy to weigh, and allowing it to vaporize at room temperature.

Procedure

This experiment is done in pairs. Each group should obtain a butane gas cylinder with hose fitting, a baby bottle with a glass plate, and a pneumatic trough. Determine the mass of the gas cylinder without the hose fitting attached and record this value on your data sheet. Fill the baby bottle completely with water and slide the glass plate over the top to keep the water in. Fill the water trough with tap water so that it is half full.

While holding the glass plate on the baby bottle, invert the bottle and immerse it in the water-filled trough. Slide the glass plate off the bottle. There should not be any air bubbles in the bottle. If there are, repeat the procedure until you can get the bottle in the trough with no air bubbles.

Run the hose from the gas cylinder into the pneumatic trough and up inside the baby bottle. It is important to **keep the butane cylinder vertical**. To dispense butane gas, depress the valve stem of the cylinder **downward**. (Do not press the valve stem sideways or the stem will break.) Collect approximately 200 mL of butane gas in the baby bottle using the graduations on the side of the bottle to measure the gas volume. The baby bottle is graduated in units of cc (cubic centimeters), which is equivalent to mL. Move the baby bottle up or down in the trough in order to get the water level in the bottle even with the water level in the trough (this ensures that the total pressure of gas in the baby bottle is equal to atmospheric pressure). Record this value on your data sheet as the volume of gas dispensed. Remove the hose from the baby bottle and tilt the bottle to let gas out to the atmosphere. Dry the hose and cylinder and reweigh the gas cylinder using the same balance as your initial weighing. The difference between mass readings corresponds to the mass of gas collected in the baby bottle.

The total pressure of gas in the bottle is equal to the atmospheric pressure. However, because the gas was collected over water, there is some water vapor mixed with the butane gas you collected. **Dalton's Law of Partial Pressures** states that the total pressure of gas in a container is the sum of the individual pressures (partial pressures) of any gases present. For this experiment we can write:

$$P_{total} = P_{water} + P_{butane}$$

The partial pressure of water vapor at various temperatures is listed in a table on your data sheet. Your instructor will provide you with the value of the atmospheric pressure. The temperature of the gas in the baby bottle is assumed to be the temperature of the water in the trough.

Safety Precautions

Butane gas is flammable. Do not have any open flames in the laboratory during this experiment.

Waste Management

Return all materials (baby bottle, glass plates, water trough, butane cylinders, and hoses) to the front lab bench.

Molecular Weight Determination of Butane Gas Prelab Questions

Name _____ **Section** _____

1. The stoppered flask shown below contains butane gas (dark circles) and water vapor (light circles). The total pressure in the flask is 1.20 atm. (a) What is the partial pressure of butane gas in the flask? (b) What is the partial pressure of water vapor in the flask? Show your calculations.

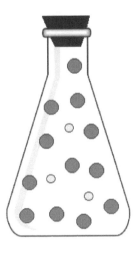

2. The diagram below represents a test tube inverted in a container of water. Is the pressure of gas in the test tube above, below, or equal to atmospheric pressure? Explain.

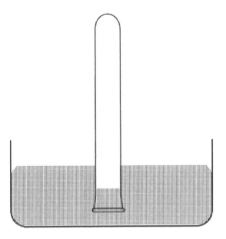

Molecular Weight Determination of Butane Gas Data Sheet

Name _____ **Section** _____

Name of Lab Partner(s) _____

1. Mass of butane cylinder initial   _____
2. Mass of butane cylinder final   _____
3. Mass of butane released   _____
4. Volume of butane collected   _____
5. Atmospheric pressure   _____
6. Partial Pressure of H_2O _____
7. Pressure of butane gas   _____
8. Temperature of butane   _____
9. Molecular weight of butane
   calculated from experiment   _____

Calculations

Vapor Pressure of Water at Various Temperatures

Temperature (°C)	Pressure (mmHg)
10	9.2
11	9.8
12	10.5
13	11.2
14	12.0
15	12.8
16	13.6
17	14.5
18	15.5
19	16.5
20	17.5
21	18.7
22	19.8
23	21.1
24	22.4
25	23.8

Molecular Weight Determination of Butane Gas Review Questions

Name _____ **Section** _____

1. Calculate the molecular weight of butane by adding the atomic weights of carbon and hydrogen (C_4H_{10}).

2. How does your experimentally determined value for the molecular weight of butane compare with the true value determined by adding the atomic weight of C and H in question 1? Although this method of determining molecular weight of a gas is a very good one, there are many sources of error inherent in this method. In addition to some of the errors your Instructor discussed, please describe at least 4 more possible sources of error as to why the molecular weight values might differ.

3. A newly discovered gas has a density of 2.39 g/L at 23 °C and 715 mmHg. What is the molecular weight of the gas?

4. Acetylene gas, $C_2H_{2(g)}$, can be prepared by the reaction of calcium carbide with water:

$$CaC_{2(s)} + 2\,H_2O_{(l)} \longrightarrow Ca(OH)_{2(s)} + C_2H_{2(g)}$$

Calculate the volume of C_2H_2 that is collected over water at 23 °C by reaction of 0.1580 g of CaC_2 if the total pressure of the gas is 726 torr? (The vapor pressure of water is listed on the data sheet).

Molecular Weight Determination of Butane Gas Supplemental Questions

Name _____ **Section** _____

pH Study

Introduction

Acid and base reactions play a key role in the chemistry of virtually all aqueous solutions. Living organisms have evolved complex mechanisms for regulating pH as slight variations in the pH of fluids such as blood can have catastrophic consequences to an organism. Likewise, large changes in the acid content of rivers and streams can have disastrous consequences on the organisms that rely on them to supply water and nutrients. In this laboratory exercise, the pH of various solutions will be measured and the relationship between the concentration of various solutes and pH will be studied.

The Concept of pH

In aqueous solution, the concentration of the hydronium ion, H_3O^+, can vary over an extremely wide range (approximately 10^1 to 10^{-15})[1]. To make it easier to deal with numbers of this size, a Danish chemist by the name of Sørensen developed the pH scale to express the concentration of hydronium ion in aqueous solutions. The term pH is defined as

$$pH = -\log [H_3O^+]$$

The brackets indicate equilibrium concentrations in units of moles/liter (molarity). By measuring the pH of a solution, we are measuring the H_3O^+ concentration of the solution. To obtain hydronium ion concentrations from pH values, we use the following relationship.

$$[H_3O^+] = 10^{-pH}$$

Water undergoes **autoionization** to produce H_3O^+ and OH^- according to the following equation.

$$2\,H_2O \rightleftharpoons H_3O^+_{(aq)} + OH^-_{(aq)}$$

In aqueous solutions at 25 °C, the product of the concentrations of H^+ and OH^- is 1.0×10^{-14}. This is expressed in the following equation where K_w is the **ionization constant for water**.

$$[H_3O^+][OH^-] = K_w = 1.0 \times 10^{-14}$$

---

[1]For convenience, we will write the aqueous hydrogen ion as H_3O^+. This is the ion that is usually referred to as the hydronium ion. However, it is important to understand that a proton in water is highly solvated and probably exists as many different species such as H_3O^+, $H_5O_2^+$, $H_9O_4^+$, etc. No single representation of the hydronium ion is adequate for all purposes, but H_3O^+ will suffice for now.

Taking the logarithm of both sides of this equation leads to the very useful relationship

pH + pOH = 14

In *pure* water, the only source of hydronium and hydroxide ions is the autoionization reaction. Because this reaction always produces equal numbers of hydroxide and hydronium ions, the concentrations of H_3O^+ and OH^- must be equal. In pure water at 25 °C, these concentrations are both 1.0×10^{-7}. Thus, *the pH of pure water will be equal to 7.0*, and any solution that has a pH of 7 is said to be **neutral**. Solutions with pH values below 7 are said to be **acidic** and those with pH values above 7 are said to be **basic**.

One technique for measuring the pH of a solution involves the use of chemical substances called **indicators**. An indicator is any substance that undergoes an easily detectable change in a physical or chemical property (usually color) in response to a change in the pH of the solution. Most indicators change color over a fairly narrow pH range (1 to 2 units). Although indicators are very useful, they have several limitations. The color changes that indicators undergo are usually gradual and the colors produced are subject to individual interpretation. The dye used to make litmus paper is an example of an acid-base indicator.

A technique that gives better quantitative results for the concentraiton of H_3O^+ in a solution, and does not alter the chemical composition of the solution, involves an instrument known as a **pH meter**. A pH meter consists of two components; an **electrode** that senses the concentration of H_3O^+ in solution, and an **amplifier** that converts the voltage from the electrode to a pH value and displays it. The pH electrode contains a solution with a fixed concentration of H_3O^+ separated from the solution to be measured by a glass membrane. The electrode responds to the difference in H_3O^+ concentration on both sides of the membrane by producing a voltage. This voltage is detected by the amplifier and converted to a pH reading. The proper use and care of a pH meter and pH electrode are described in detail in **Appendix E**.

In this experiment you will measure the pH of solutions containing *strong* acids and bases, *weak* acids and bases, and *salts* of strong and weak acids and bases. Each of these measurements will be discussed and are summarized in Table 1.

Strong Acids and Bases

Strong acids and bases are those which *ionize (dissociate) completely* in water. Strong acids produce H_3O^+ and the anion of the acid (referred to as the conjugate base of the acid). Strong bases produce OH^- and the cation of the base (referred to as the conjugate acid of the base). In solutions of monoprotic strong acids, since the acid is completely ionized, the pH will be equal to the negative log of the analytical concentration of the acid. For example, the pH of a 0.010 M solution of HCl is –log(0.010) or 2. For solutions of strong bases, it is usually easier to first calculate the pOH, and then convert pOH to pH. For example, the pOH of a 0.010 M solution of NaOH will be –log(0.010) or 2. The pH is found by subtracting the pOH from 14, giving a pH value of 12. The following expressions can be used to calculate the pH of a strong acid or base solution.

For strong acids: pH = – log C_A (C_A is the analytical concentration of the acid)

For strong bases: pH = 14 – (–log C_B)   (C_B is the analytical concentration of the base)

Examples of the ionizations of strong acids and strong bases in water are shown below.

$$HCl_{(g)} + H_2O_{(l)} \longrightarrow H_3O^+_{(aq)} + Cl^-_{(aq)} \qquad \text{strong acid}$$

$$NaOH_{(s)} \longrightarrow Na^+_{(aq)} + OH^-_{(aq)} \qquad \text{strong base}$$

The seven common strong acids include: HNO_3, H_2SO_4, $HClO_3$, $HClO_4$, HCl, HBr, and HI. All the Group IA and IIA hydroxides (such as NaOH and KOH) are considered strong bases with the exception of Be and Mg hydroxides.

Weak Acids and Bases

Weak acids and bases *do not completely ionize* in aqueous solution. In fact, most are less than 1% ionized in aqueous solution. Therefore, *the pH of a solution of weak acid will always be greater than the negative log of the analytical concentration of the acid and the pH of a solution of weak base will always be less than the negative log of the analytical concentration of the base.*

The ionization reactions of weak acids and weak bases are equilibrium processes because weak acids and bases are only partially ionized in solution. The equilibrium constants K_a (for weak acids) and K_b (for weak bases) are used to describe the extent to which a weak acid or base ionizes. The equilibrium constant value tells how much the acid or base ionizes in solution. The larger the value of the equilibrium constant for a given acid or base indicates the *stronger* that acid or base. In this experiment you will determine the K_a, called the acid dissociation constant, for the acetic acid and make some comparisons of K_a with varying the concentration of the acid.

Examples of the ionization of weak acids and bases and their corresponding equilibrium expressions, K_a and K_b, are shown below. (Note that water is omitted in the equilibrium expression because it acts as solvent in the system.)

weak acid: $\qquad HF_{(g)} + H_2O_{(l)} \rightleftharpoons H_3O^+_{(aq)} + F^-_{(aq)}$ $\qquad\qquad$ $pH > (-\log C_A)$

$$K_a = \frac{[H^+]\,[F^-]}{[HF]}$$

weak base: $\qquad NH_{3(g)} + H_2O_{(l)} \rightleftharpoons NH_4^+_{(aq)} + OH^-_{(aq)}$ $\qquad\qquad$ $pH < (-\log C_B)$

$$K_b = \frac{[NH_4^+]\,[OH^-]}{[NH_3]}$$

Reactions of Salts with Water

Aqueous solutions of salts (ionic compounds) can be acid, basic, or neutral depending on whether or not the ions in the salt react with water. The reactions of ions with water to produce acidic or basic solutions is called **hydrolysis.** The cations of Group 1A and 2A do not react with water and are therefore neutral in aqueous solution. Transition metal cations, on the other hand, are almost all acidic in aqueous solution (e.q. Fe^{3+}, Cu^{2+}, Ag^+, Zn^{2+}). Another common acidic cation is the ammonium ion, NH_4^+. NH_4^+ is the conjugate acid of the weak base ammonia, NH_3, and is therefore acidic in water. Anions which are the conjugate bases of strong acids (e.g. Cl^-, Br^-, I^-, NO_3^-) are all neutral in water. Conversely, anions which are the conjugate bases of weak acids (e.g. F^-, NO_2^-, CN^-, OAc^-) are all basic in aqueous solution. Shown below are a few examples of hydrolysis reactions.

Conjugate Base of Weak Acid:

$\qquad CN^- + H_2O_{(l)} \rightleftharpoons HCN_{(aq)} + OH^-_{(aq)}$ $\qquad\qquad$ $pH > 7$

$\qquad OAc^- + H_2O_{(l)} \rightleftharpoons HOAc_{(aq)} + OH^-_{(aq)}$ $\qquad\qquad$ $pH > 7$

Conjugate Acid of Weak Base:

$\qquad NH_4^+ + H_2O_{(l)} \rightleftharpoons NH_{3(aq)} + H_3O^+_{(aq)}$ $\qquad\qquad$ $pH < 7$

Transition Metal Cation:

$\qquad Fe^{3+} + 6\,H_2O_{(l)} \rightleftharpoons Fe(OH)(H_2O)_5^{2+}_{(aq)} + H_3O^+_{(aq)}$ $\qquad\qquad$ $pH < 7$

All of these interactions with water to produce acidic, basic, or neutral solutions is summarized in Table 1.

Table 1 *Reactions of Strong and Weak Acids and Bases and Their Salts with Water*

Substance	Type	Example Reactions in Aqueous Solution	Prediction of pH
Acid	Strong	$HCl + H_2O \longrightarrow H_3O^+ + Cl^-$	$pH = -\log C_A$
Acid	Weak	$HF + H_2O \rightleftharpoons H_3O^+ + F^-$	$pH > (-\log C_A)$
Base	Strong	$NaOH \longrightarrow Na^+ + OH^-$	$pH = 14 - (-\log C_B)$
Base	Weak	$NH_3 + H_2O \rightleftharpoons NH_4^+ + OH^-$	$pH < 14 - (-\log C_B)$
Conjugate acid of strong base	Neutral	No hydrolysis reaction occurs	$pH = 7$
Conjugate base of strong acid	Neutral	No hydrolysis reaction occurs	$pH = 7$
Conjugate base of weak acid	Weak	$F^- + H_2O \rightleftharpoons HF + OH^-$	$pH > 7$
Conjugate acid of weak base	Weak	$NH_4^+ + H_2O \rightleftharpoons H_3O^+ + NH_3$	$pH < 7$
Transition metal cation	Weak	$Fe^{3+} + 6H_2O \rightleftharpoons Fe(OH)(H2O)_5^{2+} + H_3O^+$	$pH < 7$

Buffers

A **buffer solution** is an aqueous solution containing a weak acid and its conjugate base (an acidic buffer) or a weak base and its conjugate acid (a basic buffer). A buffer solution has the ability to resist changes in pH upon the addition of small amounts of acid or base. The acid in the buffer reacts with any extra OH^- while the base in the buffer reacts with any extra H_3O^+. You will have the opportunity to study the properties of a buffer solution using an acetic acid (CH_3COOH which can be abbreviated as HAc or HOAc) and sodium acetate ($NaCH_3COO^-$, which can be abbreciated as NaOAc) buffer system.

Procedure

Part A: Acids and Bases

Before starting today's experiment, review the material in **Appendix E** on the proper use and care of pH electrodes. Your instructor will assign you three of the six unknown solutions used in this part of the experiment. You will know the concentrations of the unknowns, but you will not know their identities (weak acid, strong acid, etc.). Record the concentrations of your unknown solutions on your data sheet.

Using the concentrations reported on the labels, calculate what the pH of each unknown solution *should* be if it contained a strong acid or a strong base. For example, if the concentration listed on the label of one of your unknowns was 0.05 M, the calculations would look like this.

If the unknown is a strong acid: pH = − log(0.05) = 1.3

If the unknown is a strong base: pH = 14 − (−log(0.05)) = 12.7

Pour 20–25 mL of your unknown solution into a 50 mL beaker and place the pH probe in the solution. Record the measured pH value in your data sheet. Compare the experimentally determined pH with the predicted pH and determine if your unknown is a strong acid, weak acid, strong base, or weak base.

Calculate the hydronium ion concentration (in moles/L) in each unknown from the pH values you measured and record these values on your data sheet.

Part B: Reactions of Salts with Water

Your instructor will assign you three of five unknown solid salts. The molar mass of each salt is written on its container. Prepare a 0.10 M solution of each of your unknown salts by weighing 0.0020 mole of the solid and adding it to 20.0 mL of DI water. Record the color of solids and solutions prepared. Determine the pH of each unknown salt solution and record these values on your data sheet.

Based on the pH you measured for each of your unknown solutions, the molar masses of the salts and colors in solution, you may identify each unknown solid from the list of possible compounds below. In the space provided on your data sheet, list at least 2 compounds from the following list which will produce the acidic, basic, or neutral solutions you observed for each unknown salt reaction with water. Lastly, circle the ion responsible for producing the acidic or basic solution.

Possible Unknown Salts:

NH_4Cl, Li_2CO_3, $Ca(NO_3)_2$, KCl, $Mg(NO_3)_2$, Na_2CO_3, CaF_2, $Zn(NO_3)_2$, $CuSO_4$,

$NaCl$, KNO_3, LiF, NH_4Br, $CaCO_3$, NaF, $Pb(NO_3)_2$, $NaOAc$, $LiCl$, $BeCl_2$

Part C: K_a Acid Dissociation Constant

Determine the pH of the three different solutions of acetic acid (CH_3COOH which is sometimes abbreviated as HOAc, or HAc): 1.0 M, 0.10 M, and 0.010 M. Enter the data on your data sheet. From the pH determine the [H^+] at each of these concentrations and enter these values on your data sheet. Write the equilibrium expressions and calculate K_a, the acid dissociation constant, for acetic acid for each of these concentrations. The equilibrium for a weak acid, HX, is as follows:

$$K_a = \frac{[H^+][X^-]}{[HX]}$$

Because H^+ and the acetate anion, Ac^- are in a 1:1 ration in the dissociation reaction of acetic acid, the AC^- concentration will be equal to the H^+ concentration. The concentration of HX in the equilibrium expression (undissociated acetic acid, CH_3COOH, abbreciated as HOAC or HAc) is the initial concetnration of weak acid solution.

Part D: Buffers

Pour 20 mL of HOAc/NaOAc (acetic acid / sodium acetate buffer) solution in a 50 mL beaker and determine the pH. Record the pH on your data sheet. Add 3 drops of 1.0 M HCl to the beaker, stir, and measure the pH again. Record this value on your data sheet. To a second beaker of the buffer solution, add 3 drops of 1.0 M NaOH. Measure the pH of this solution and enter the value on your data sheet.

Add 20 mL of deionized water to a 50 mL beaker and determine its pH. Follow the same procedure you used with the buffer solution, adding HCl and NaOH to fresh samples of DI water and recording the pH. You should observe a *very* large difference in behavior between the buffer solution and pure water when an acid or base is added to it. On your data sheet write a paragraph or two discussing and comparing the use of buffer versus a non-buffered solution (DI water) with the addition of a strong acid or base.

Safety Precautions

Safety goggles must be worn throughout entire laboratory period, even when cleaning glassware to prevent damage to your eyes from strong acids and bases! Be cautious when using unknown solutions as they may be strong acids or bases. Be cautious wafting to determine odors of solutions. Strong acids and bases will cause severe burns to skin and eyes, and create holes in clothes and shoes. Clean any spills immediately with paper towels and water.

Waste Management

All waste may be flushed down the sink with plenty of water.

pH Study Prelab Questions

Name _____ **Section** _____

1. The pH of three solutions of 0.10 M HCN, HF, and HOBr are 5.1, 2.1 and 4.7 respectively. Designate which is the strongest and which is the weakest acid.

2. The pH of seawater is 8.30. What is the concentration (molarity) of $[H_3O^+]$ in seawater?

3. Predict whether reaction of each of the following compounds with water will produce an acidic, basic, or neutral solution. Use the information in the introduction to this experiment and in Table 1 to make these determinations.

    NH_4Cl

    Na_2CO_3

    $CuSO_4$

    $NaCl$

    KNO_3

    LiF

4. Briefly summarize the steps necessary for the proper care of a pH electrode (refer to **Appendix E**).

pH Study Data Sheet

Name _____ **Section** _____

Name of Lab Partner(s) _____

Part A: Acids and Bases

Action	Unknowns		
Unknown number			
Concentration of unknown shown on label			
Predicted pH of unknown if it is a strong acid			
Predicted pH of unknown if it is a strong base			
Measured pH of unknown			
Indicate if the unknown is a strong acid, weak acid, strong base, or weak base			
Concentration of H_3O^+ from measured pH			

Part B: Reactions of Salts with Water

Action	Unknowns		
Unknown number			
Molar mass of unknown solid shown on label			
Mass (g) of unknown necessary to make a 0.10 M solution			
Color of unknown solid			
Color of unknown solution			
Measured pH of 0.10 M solution of unknown solid			
Designate if Acidic, Basic, or Neutral solution produced by hydrolysis of unknown solid			
Using the list of possibilities from Part B of Procedure "Reactions of Salts with Water", list as many unknown compounds that may fit the designation of Acidic, Basic, or Neutral. In each compound, circle the ion responsible for making the solution Acidic or Basic.			

Part C: K_a, Acid Dissociation Constant

Acetic Acid Concentration	pH	[H+]	K_a Calculation	K_a
0.010 M				
0.10 M				
1.0 M				

Average K_a for acetic acid: _____

Part D: Buffers

pH of acetic acid/sodium acetate buffer solution (HOAc/NaOAc)   _____

pH of acetic acid/sodium acetate buffer solution + 3 drops of 1 M HCl   _____

pH of acetic acid/sodium acetate buffer solution + 3 drops of 1 M NaOH   _____

pH of deionized water   _____

pH of water + 3 drops of 1 M HCl   _____

pH of water + 3 drops of 1 M NaOH   _____

In the space below, discuss your observations and compare the use of a buffered versus a non-buffered solution (DI water) with the addition of a strong acid or base.

pH Study Review Questions

Name _____ Section _____

1. What is the pH and pOH of a 0.20 M NaOH solution?

2. The hydronium ion $[H_3O^+]$ concentration in a vinegar sample has been determined to be 1.6×10^{-3} M. Calculate the pH.

3. Predict whether aqueous solutions of the following compounds will be **acidic**, **basic** or **neutral**. If the solution would be acidic or basic, circle the ion that causes the pH to change.

    a) NH_4Br b) $FeCl_3$ c) K_2CO_3 d) $KClO_4$

    e) $NaBr$ f) $MgCl_2$ g) NaF h) $LiNO_3$

4. The pH of a 0.20 M proprionic acid (CH_3CH_2COOH) solution is 2.79. Calculate the K_a of proprionic acid.

5. Two unknown acid solutions are both labelled 0.50 M. How could you determine which is the strong acid and which is the weak acid? Explain.

pH Study Supplemental Questions

Name _____ **Section** _____

Polymers: Nylon Synthesis

Introduction

Polymers, sometimes called **resins**, are the main products of the chemical industry. They furnish the major constituents for all plastics and fibers. Polymers are very large molecules and are made by adding many small molecules called **monomers**, together to form long chains. Some of the best known types of polymers are nylons, polyester, acrylics, polyvinyls, and polystyrenes. Nylons, polyesters, and acrylics are used mainly in the clothing industry. Polyvinyls are used to make plastic sheeting and plumbing materials. Polystyrenes are used extensively for insulation materials. Styrofoam cups are made from a type of polystyrene.

In today's experiment you will be making a polymer known as **nylon 6, 10**. Nylon is an example of a **polyamide**. An amide is made by chemically combining an organic acid chloride with an amine. An equation showing the reaction of an acid chloride and an amine to make an amide is shown in Figure 1 below.

To make a polyamide, it is necessary for the amine molecule to contain an $-NH_2$ group at each end (this is known as a diamine) and for the acid chloride molecule to have a $-COCl$ group at each end (a diacid chloride). The diamine and the diacid chloride can then bond together, end-on-end, to form very long chains. Nylon 6,10 is made from a diamine having six carbon atoms (hexamethyenediamine) and a diacid chloride having ten carbon atoms (sebacoyl chloride). The formulas for these two compounds are shown in Figure 2 below.

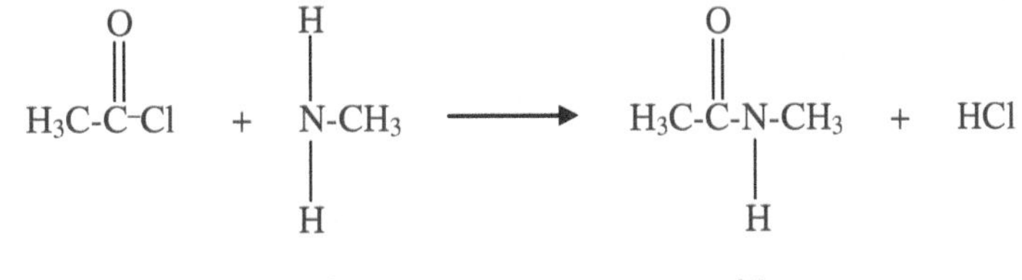

$$H_3C-\overset{\overset{\displaystyle O}{\|}}{C}-Cl \quad + \quad \overset{\overset{\displaystyle H}{|}}{\underset{\underset{\displaystyle H}{|}}{N}}-CH_3 \quad \longrightarrow \quad H_3C-\overset{\overset{\displaystyle O}{\|}}{C}-\underset{\underset{\displaystyle H}{|}}{N}-CH_3 \quad + \quad HCl$$

acid chloride   **amine**   **amide**

FIGURE 1 — *Synthesis of an Amide*

$$H_2N-(CH_2)_6-NH_2 \qquad\qquad\qquad Cl-\overset{\overset{\displaystyle O}{\|}}{C}-(CH_2)_8-\overset{\overset{\displaystyle O}{\|}}{C}-Cl$$

hexamethylenediamine   **sebacoyl chloride**

FIGURE 2 — *Structures of Hexamethylenediamine and Sebacoyl Chloride*

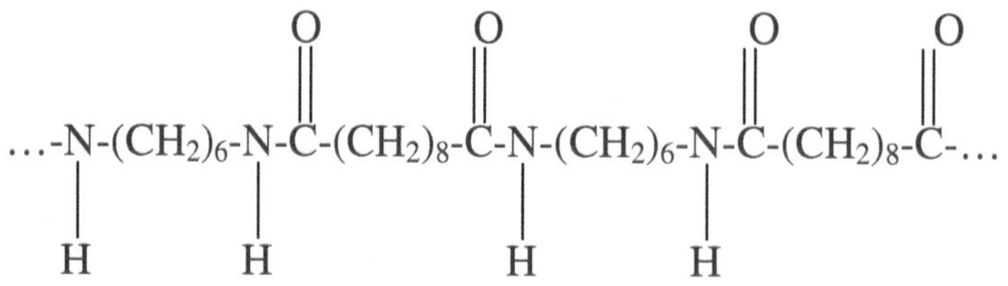

FIGURE 3 — *Nylon 6,10*

When the diacid chloride and the diamine are added together they polymerize to form nylon 6,10. A portion of a nylon 6,10 molecule is shown in Figure 3 above.

In the second part of this experiment you will compare the properties of nylon 6,10 against those of polystyrene. A portion of a polystyrene molecule is shown in the diagram in Figure 4.

You should note that polystyrene contains only carbon and hydrogen atoms while nylon contains carbon, hydrogen, nitrogen and oxygen atoms. This will make nylon more polar than polystyrene and also give nylon the capacity for forming intermolecular hydrogen bonds.

In characterizing polymers, it is important to know something of both their physical and chemical properties. The **physical properties** of a polymer, such as melting point and solubility, depend mostly on the **intermolecular forces** in the polymer. The **chemical reactivity** of a polymer depends on the types of atoms in the molecule and the **intramolecular bonding** between them (single bonds, double bonds, etc). A polymer may be quite stable with respect to one kind of reaction and unstable with respect to another. A frequently encountered reaction of organic compounds that contain nitrogen is air oxidation, which causes them to discolor. In some cases this occurs at room temperature, while with other substances it is a problem only at high temperatures.

Procedure

Part A: Synthesis of Nylon 6,10

The stock solutions contain hexamethylenediamine and NaOH dissolved in water (labeled "diamine"), and sebacoyl chloride dissolved in hexanes (labeled "diacid chloride"). Pour 5 mL of the diamine into a 50 mL beaker. You may use the graduations on the side of the beaker to measure the 5 mL. Pour 10 mL of the diacid chloride into a graduated cylinder.

Slowly pour the diacid chloride down the side of the beaker containing the diamine. A white film should form at the interface of the two layers. Allow the chemicals to rest a few minutes undisturbed while the reaction takes place at the interface of the two liquids.

Reach into the beaker with forceps and grasp the film in the center of the beaker. Slowly pull straight up. Be careful not to let the thread of nylon touch the sides of the beaker. Pull the nylon from the beaker and wrap the thread around your largest test tube. Rotate the tube, counting revolutions, until no more nylon can be obtained. If the thread breaks at any time, grasp the file with the forceps and continue. Record the number of revolutions of nylon on the data sheet.

Wash the nylon with water and then with acetone to hasten the drying process. Press the washed nylon between paper towels until it no longer makes them wet. Weigh the nylon and enter the mass on your data sheet.

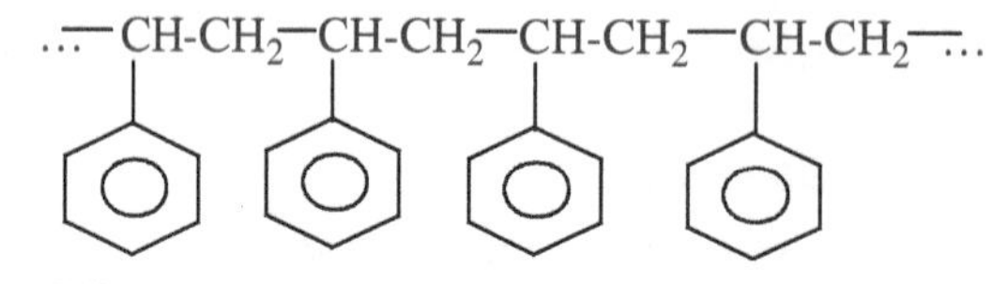

FIGURE 4 — *Polystyrene*

Part B: Properties of Polymers

In this part of the experiment, you will compare the properties of foamed polystyrene with those of nylon 6,10. Place a small piece of polystyrene on the tip of your metal spatula. Heat it in your Bunsen burner flame and note how long it takes to melt and whether or not it changes color. Repeat the process with the same size piece of nylon.

Place 3 mL of toluene (C_7H_8, a nonpolar solvent) in each of two medium sized test tubes. Add a small piece of nylon to one of the tubes and a similar sized piece of polystyrene to the other. Stopper the tubes and shake them (do not use your thumb as a stopper). Observe the tubes closely to see which polymer is the more soluble in toluene.

Safety Precautions

Caution: The reagents used in this experiment are toxic. They should be handled with care. The reaction should be carried out near your exhaust hood. If you come in contact with any of the reagents, immediately rinse the affected area with plenty of water.

Waste Management

Pour the toluene into the "Organics waste" bottle in the fume hood. Nylon may go in the wastebasket.

Polymers: Nylon Synthesis
Prelab Questions

Name _____ **Section** _____

1. Place a circle around a monomer unit in the <u>polystyrene</u> drawing in the Introduction section of this laboratory experiment and draw the monomer below.

2. Explain why the two solutions, sebacoyl chloride and hexamethylenediamine, are not soluble in one another.

3. Considering the chemical structure and intermolecular forces of the two polymers, nylon or polystyrene, which should have the higher melting point?

4. Name several polymers that occur in nature. (Example: silk from silk worm).

Polymers: Nylon Synthesis Data Sheet and Review Questions

Name _____ **Section** _____

Name of Lab Partner(s) _____

Data

Number of revolutions _____

Mass of nylon produced _____

Questions

1. Your large test tube has a diameter of 25 mm. Knowing the number of revolutions used in obtaining your nylon, calculate the length (in feet) of the nylon thread you made. (Hint: $C = \pi d$, where C is circumference, d is the diameter of the circle, and $\pi = 3.14$.) Show your calculations below.

2. If the total cost of the materials used to make your nylon is $0.55, what would a pound of your nylon cost? Show your calculations.

3. Which polymer had the lower melting point? _____ Explain the difference in melting points in terms of the intermolecular attractive forces in each of the polymers.

4. Which polymer darkened upon heating? _____ Based on this observation, which polymer is less stable toward air oxidation? Explain this in terms of polymer structure.

5. Which polymer is the most soluble in toluene? _____ Explain the results of the solubility experiment in terms of the intermolecular attractions between polymer molecules and solvent molecules.

Polymers: Nylon Synthesis
Supplemental Questions

Name _____ Section _____

Qualitative Analysis: Group I

Introduction to Qualitative Analysis

The term chemical analysis implies the separation of substances in a sample of unknown composition into its individual components by using chemical and/or physical separation methods. When examining a sample, the first step involves detecting and identifying individual components. **Qualitative analysis** deals with identifying the individual substances present in the sample while **quantitative analysis** is used to determine the amount of each substance in the sample.

In today's experiment, you will learn to identify the presence of certain ions in a solution using **classical qualitative analysis.** The procedure is called "classical" because it was developed and used before the advent of modern chemical instrumentation. These methods continue to be valuable tools for learning the unique chemical and physical properties of ions.

Qualitative chemical analysis relies on the fact that certain ions have very <u>similar</u> chemical reactivities toward a particular reagent and can therefore be grouped together in an analysis scheme. For example, the ions that comprise Group I (Ag^+, Pb^{2+}, and Hg_2^{2+}) all form insoluble chlorides when mixed with HCl[1]. The identification of individual ions in the scheme is based on the <u>differences</u> in their reactivities toward other reagents. These differences include such things as the formation of precipitates, reactions with acids and/or bases, and the formation of colored complexes.

The procedures for separating groups of ions from each other and for separating ions within groups are often presented in an outline form called a **flowchart**. A flowchart consists of a series of branches called *separation steps* that show how ions will be separated from each other. Each branch eventually terminates in a *confirmatory test* that unambiguously established the presence or absence of that ion in the sample.

Before beginning a description of the methods for the Group I separation and identification, several technical points need to be emphasized.

1. The use of clean containers (beakers, test tubes, spot plates, stirring rods, pipettes, etc.) is <u>very</u> important. Thoroughly wash all glassware, followed by a rinse with deionized (DI) water after each use. Dirty containers and tap water contain traces of interfering ions that may lead to erroneous conclusions during your analysis.

2. Be careful to use the reagent specified in the test you are conducting. There will be several different concentrations of each acid and base present in the lab and it is easy to grab the wrong one. Check the label before using any reagents.

---

[1]Qualitative analysis Group I should not be confused with Group 1A of the periodic table.

3. When describing the results of tests, be sure to distinguish among clear, cloudy, colored, and colorless for solutions and mixtures. An insoluble precipitate may resemble a cloudy, turbid, or gelatinous substance in addition to the typical distinct look of a crystalline solid. When performing a test, always mix reagents thoroughly with a clean stirring rod.

4. Some of the procedures call for **centrifuging** a sample. A centrifuge uses centrifugal force to separate the liquid and solid portions of a mixture. (Centrifugal force causes rotating bodies to be moved away from the center of rotation.) The centrifuges are located on the benches in the front and back of the lab room. Always balance the centrifuge by placing a **counterbalance** test tube filled with water to about the same level as your sample directly across from your sample test tube before centrifuging. This keeps the centrifuge from vibrating and maintains consistent separation of your sample.

5. The liquid above the precipitate in your sample test tube after centrifugation is called the **supernatant.** Pay careful attention to the procedure because in several analyses you will use the supernatant for a subsequent step. The best way to remove the supernatant without disturbing the solid precipitate is to withdraw the liquid by using a disposable or capillary pipet. **Decanting** can also be performed by carefully pouring off the liquid portion into another test tube.

6. The precipitate separated from the supernatant may still contain a small amount of liquid on the surface; therefore, it is necessary to *wash* the precipitate to remove any contaminating ions from a previous separation step. Washing the precipitate is accomplished by adding a small amount of deionized water to the precipitate. After stirring, the mixture is centrifuged and decanted to remove the wash water

7. When adding an acid or base to a mixture, always ensure complete reaction by stirring well with a clean stirring rod. Withdraw a drop of the mixture with your stirring rod and apply the drop to a strip of litmus paper to determine if the mixture is acidic or basic. To avoid contamination, do not dip the litmus paper directly into the mixture.

8. Use a **water bath** to heat all mixtures. Do not heat any of your samples directly in the burner flame. When liquids contained in a micro test tube are heated directly in a flame, they have a tendency to **bump** (shoot out of the test tube under pressure).

Introduction to Group I Analysis

The cations of Group I (Ag^+, Pb^{2+}, and Hg_2^{2+}) all form precipitates with the chloride ion at room temperature resulting in the formation of insoluble $AgCl$, $PbCl_2$, and Hg_2Cl_2. (Note: Hg_2^{2+} is the mercurous ion – mercury(I) – and is an example of a diatomic ion.) Because ions in a qualitative analysis scheme are separated on the basis of the solubilities of compounds they form, it is possible to separate Group I ions from a mixture of other ions by adding chloride ions to the solution.

Lead chloride is insoluble in cold water but dissolves in *hot* water. Lead chloride can therefore be separated from silver(I) and mercury(I) chlorides by raising the temperature of the mixture. Silver chloride can be separated from mercury(I) chloride by the addition of aqueous ammonia ($NH_{3(aq)}$, often written as NH_4OH). Silver ions form a soluble **complex ion** with ammonia molecules called diammine silver(I), $Ag(NH_3)_2^+$, while mercury(I) ions precipitate as a grayish mixture of $HgNH_2Cl$ and metallic mercury. All these separations and confirmations are summarized in the flow chart shown in Figure 1.

Procedure

You should perform qualitative analysis tests on a **known** and an **unknown** sample simultaneously so you can directly compare the results obtained with the known sample with those for your unknown. Record all your observations for both the known and unknown on your data sheet.

Step 1. Prepare a Group I **known** by mixing 8 drops each of the nitrate solutions of Ag^+, Hg_2^{2+}, and Pb^{2+} in a micro (small) test tube. Label this test tube "**K**" to indicate it is your (**known**) mixture containing all three

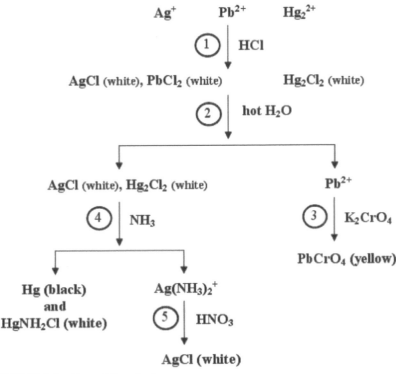

FIGURE 1 — *Group I Flowchart*

Group I cations. You will be assigned an **unknown** solution that contains one or more ions from Group I. Be sure to record the unknown number on your data sheet. Label all test tubes used in the unknown determination with the symbol "U" for "**unknown**". Obtain approximately 20–25 drops of unknown solution in a micro test tube.

Add 2 drops of 6 M HCl to the solution in each test tube. A white precipitate will form due to the formation of the chlorides of Group I cations. The **net ionic equations** for these reactions follow. Net ionic equations are balanced equations that contain only the ions that chemically react. **Spectator ions,** which are present in the reaction mixture, but do not participate in the chemical reaction and are not included in the balanced equation (such as NO_3^- and H^+ in the following reactions).

$$Ag^+ + Cl^- \longrightarrow AgCl_{(s)} \text{ (white)}$$

$$Pb^{2+} + 2 Cl^- \longrightarrow PbCl_{2(s)} \text{ (white)}$$

$$Hg_2^{2+} + 2 Cl^- \longrightarrow Hg_2Cl_{2(s)} \text{ (white)}$$

Counterbalance the known test tube with your unknown test tube and centrifuge the mixture for about 30 seconds. After the separation of the supernatant and precipitate is complete, remove both test tubes from the centrifuge and add one more drop of 6 M HCl to test for complete precipitation. If more precipitate forms, centrifuge both test tubes again. Decant the supernatant and wash the precipitate by adding 1 mL of deionized water followed by stirring and centrifuging. Discard the wash supernatant and save the precipitate for Step 2.

Step 2. Setup a hot water bath by placing 50 mL of water in a 100 mL beaker and bringing the water to a boil. Add 1 mL deionized water to the test tubes containing the precipitates and stir. Place the test tubes in the boiling water bath for 3 minutes, stirring every 30 seconds to dissolve any $PbCl_2$ that may be present. Be careful not to let any of the water from the boiling water bath splash over into the test tubes. While still hot, remove the test tubes from the water bath, centrifuge quickly, and decant the supernatants into clean micro test tubes. The supernatant will be used in Step 3. The precipitate will be used in Step 4.

$$\text{PbCl}_{2(s)} \xrightarrow{\text{heat}} \text{Pb}^{2+}_{(aq)} + 2\,\text{Cl}^-_{(aq)}$$

Step 3. To the supernatant from Step 2, add 3 drops of 1 M K_2CrO_4. A bright yellow precipitate, $PbCrO_4(s)$, confirms the presence of Pb^{2+} ions.

$$\text{Pb}^{2+} + \text{CrO}_4^{2-} \longrightarrow \text{PbCrO}_{4(s)} \text{ (yellow)}$$

Step 4. The precipitate from Step 2 should be washed with 1 M NaOAc (sodium acetate) to remove any Pb^{2+} before continuing. Add 5–10 drops 1 M NaOAc, stir, centrifuge, and discard the wash supernatant.

Add 10 drops of 6 M NH_4OH (ammonium hydroxide) to the precipitate after washing. Swirl or stir to mix and centrifuge to separate the solid from the liquid. Any silver chloride precipitate present from Step 2 will readily dissolve upon the addition of ammonia to give the water soluble complex ion, $Ag(NH_3)_2^+{}_{(aq)}$.

If Hg_2^{2+} ion is present, a gray precipitate will form which consists of black metallic mercury (finely divided metals usually appear black) and a white precipitate, $HgNH_2Cl$. Decant the supernatant, which may contain aqueous $Ag(NH_3)_2^+$, and save it for Step 5.

$$\text{Ag}^+ + 2\,\text{NH}_3 \longrightarrow \text{Ag(NH}_3)_2^+{}_{(aq)}$$

$$\text{Hg}_2\text{Cl}_2 + 2\,\text{NH}_3 \longrightarrow \text{Hg}_{(l)} \text{ (black)} + \text{HgNH}_2\text{Cl}_{(s)} \text{ (white)} + \text{NH}_4\text{Cl}_{(aq)}$$

Step 5. Add 6 M HNO_3 drop-wise to the supernatant from Step 4 until it is acidic to litmus. Test for acidity by withdrawing a drop of the solution from the test tube and placing it on a piece of blue litmus paper. Make sure the solution is well stirred before each litmus test. The acid reacts with the NH_3 in the diammine silver complex ion allowing the Ag^+ to reunite with the free Cl^- ions present in the mixture. The formation of a white AgCl precipitate in the acidified solution indicates the presence of Ag^+ in the unknown.

$$\text{Ag}^+ + \text{Cl}^- \longrightarrow \text{AgCl}_{(s)} \text{ (white)}$$

Safety Precautions

Be cautious when using hydrochloric acid, nitric acid and ammonium hydroxide as these cause burns. Wash your hands thoroughly after this experiment because you worked with lead and mercury compounds. Clean up any spills immediately.

Waste Management

After you have completed the steps to confirm the presence of a cation, place the waste in the container labeled for that cation. **Be sure to use the proper container!** Unknown and known wastes from steps #3 and #5 should be poured in the "Metals waste" container. All wastes from step #4 should be rinsed into the "Mercury waste" container.

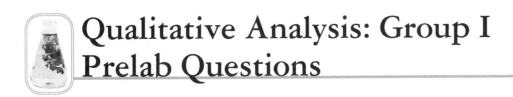

Qualitative Analysis: Group I
Prelab Questions

Name _____ **Section** _____

1. Why is it important to use deionized water in cleaning glassware between steps?

2. Why is the centrifuge used in steps where precipitates are formed?

3. How would your results be affected in Step 3 if the transferred supernatant from Step 2 of the Qualitative Analysis Group I procedure were <u>cool</u> rather than hot? Explain.

Qualitative Analysis: Group I
Data Sheet

Name _____ Section _____

Name of Lab Partner(s) _____

Group I Unknown Number _____

Step	Known	Unknown
1. Observations:		
2. Observations:		
3. Observations:		
4. Observations:		
5. Observations:		

Ions Identified in Unknown: _____ _____ _____

Qualitative Analysis: Group I
Review Questions

Name _____ Section _____

1. When silver nitrate is added to hydrobromic acid a pale yellow, curdy precipitate forms. Write the net ionic equation for this reaction.

2. (a) Write a balanced equation for the exchange reaction that occurs between $(NH_4)_2S$ and $ZnCl_2$. (b) Write the balanced net ionic equation for this reaction.

3. A student had an unknown that formed a white precipitate after the addition of HCl. The precipitate was placed in a hot water bath for several minutes and dissolved. What can the student conclude about the identity of the unknown? Explain.

4. Another student had an unknown that formed a white precipitate after the addition of HCl. After treatment in a hot water bath, the hot supernatant liquid was separated and gave no precipitate upon the addition of K_2CrO_4. The remaining precipitate was treated with ammonia and the solid dissolved leaving only a clear, colorless solution. What can the student conclude about the identity of the unknown? Explain.

Qualitative Analysis: Group I
Supplemental Questions

Name _____ **Section** _____

Qualitative Analysis: Group III

Introduction

The qualitative analysis Group III scheme comprises seven cations: Al^{3+}, Cr^{3+}, Fe^{3+}, Zn^{2+}, Ni^{2+}, Co^{2+}, and Mn^{2+}. In the traditional Group III analysis, these ions are precipitate from a basic sulfide solution as Fe_2S_3, ZnS, NiS, CoS, MnS, $Al(OH)_3$, and $Cr(OH)_3$. In today's experiment, you will be working with only four of these seven cations: Fe^{3+}, Al^{3+}, Ni^{2+}, and Cr^{3+}. These cations will be separated based on differences in the solubilities of the **hydroxides** they form. In contrast to the dense, powder-like precipitates formed in the Group I analysis, hydroxide precipitates are often gelatinous, slippery, or flaky and do not settle very well to the bottom of the test tube.

Another difference that is easily observed between the cations of Group I and Group III is that most of the Group III ions exhibit color in aqueous solution. Fe^{3+} ions are yellow, Ni^{2+} ions are green, and Cr^{3+} ions are greenish-blue to black depending on the anion present. Al^{3+} ions are colorless in aqueous solution. A quick visual inspection of your Group III unknown can give you important information about which ions may, or may not, be present.

The initial separation step in the Group III analysis involves the use of a basic hydrogen peroxide solution to oxidize Cr^{3+} to Cr^{6+} (in the form of the chromate ion, CrO_4^{2+}) and to precipitate Fe^{3+} and Ni^{2+} as their hydroxides. The Al^{3+} ion is converted to a soluble complex ion, $Al(OH)_4^-$, known as the aluminate ion. In the presence of a small amount of hydroxide ion, Al^{3+} will precipitate as $Al(OH)_3$. When the concentration of hydroxide is high, Al^{3+} forms the soluble aluminate ion, $Al(OH)_4^-$.

$$Al^{3+}_{(aq)} + OH^-_{(aq)} \longrightarrow Al(OH)_{3(s)} \quad \textbf{(low } OH^- \text{ concentration)}$$

$$Al^{3+}_{(aq)} + OH^-_{(aq)} \longrightarrow Al(OH)^-_{4\,(aq)} \quad \textbf{(high } OH^- \text{ concentration)}$$

The precipitates of iron and nickel hydroxides formed in Step 1 are subsequently dissolved in acid and the iron(III) hydroxide reprecipitated from an aqueous ammonia solution. The Ni(II) ion forms a soluble complex ion with ammonia, $Ni(NH_3)_6^{2+}$, rather than precipitating as a hydroxide. Confirmation of nickel involves complexing nickel(II) ion with dimethylglyoxime to give a bright red precipitate. The presence of iron(III) ion is confirmed by redissolving iron(III) hydroxide in an acidic solution and reacting the iron(III) ion with thiocyanate. This produces the blood red $FeSCN^{2+}$ ion.

The CrO_4^{2-} and $Al(OH)_4^-$ ions are separated from each other by controlling the pH of the solution so that aluminum ions form insoluble aluminum hydroxide, $Al(OH)_3$. The aluminum hydroxide precipitate is gelatinous and therefore very difficult to see. To confirm the presence of $Al(OH)_3$, a special dye known as aluminon is added to color the precipitate. Confirmation of chromium(III) involves precipitation with barium to form a light yellow precipitate of barium chromate.

The flowchart for the separation of the Group III ions is shown in Figure 1. The circled numbers in the flowchart refer to specific steps described in the procedure.

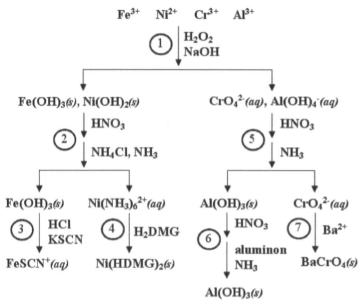

FIGURE 1 — *Group III Analysis Flowchart*

Procedure

You should perform qualitative analysis tests on a **known** and an **unknown** sample simultaneously so you can directly compare the results obtained with the known sample with those for your unknown. Record all your observations for both the known and unknown on your data sheet.

Step 1. Pay careful attention to the labels and concentrations of all reagents used in this experiment. Both nitric acid (HNO_3) and aqueous ammonia (NH_4OH) will be used in two different concentrations; 6 M and 15 M.

Prepare a Group III known by mixing 5 drops each of 0.1 M nitrate solutions of Al^{3+}, Ni^{2+}, Cr^{3+}, and Fe^{3+} in a micro test tube. Obtain approximately 20 drops of your unknown sample. From this point on the procedures that are described should be performed on both the known and unknown solutions. Add 3 drops of 3% H_2O_2 and then add 15 drops of 6 M NaOH and stir. Heat the test tube in a hot water bath to bring the contents of the test tube to a boil. This will decompose any H_2O_2 left in the tube. Centrifuge and decant the supernatant. If the solution contained chromium ions, the supernatant should be yellow in color. Save the supernatant containing aluminate and chromate ions for Step 5. Follow the procedure in Step 2 for the precipitates of iron and nickel hydroxides. Net ionic equations for the reactions in Step 1 are shown below.

$$10\ OH^- + 2\ Cr^{3+} + 3\ H_2O_2 \longrightarrow 2\ CrO_4^{2-}{}_{(aq)}\ (yellow) + 8\ H_2O$$

$$Al^{3+} + 4\ OH^- \longrightarrow Al(OH)_4^-{}_{(aq)}\ (colorless)$$

$$Fe^{3+} + 3\ OH^- \longrightarrow Fe(OH)_{3(s)}\ (rust\ color)$$

$$Ni^{2+} + 2\ OH^- \longrightarrow Ni(OH)_{2(s)}\ (green)$$

Step 2. Add 8 drops of 15 M HNO_3 to the precipitate from Step 1 and heat the mixture in a hot water bath until the precipitate has completely dissolved. Let the solution cool and then add 10 drops of 1 M NH_4Cl and 10 drops of 15 M NH_3 (NH_4OH). Test the solution with litmus paper to see if it is basic. If not, continue to add 15 M NH_3 drop-wise until the solution becomes basic. When the solution is basic, add 4 more drops of 15 M NH_3. Any iron(III) ions present will precipitate as rust-colored $Fe(OH)_3$. Ions of Ni^{2+} form the soluble complex ion $Ni(NH_3)_6^{2+}$. Centrifuge and decant the supernatant, which will be used in Step 4 for confirmation of nickel(II). Save the precipitate for use in Step 3 to confirm Fe^{3+}.

$$Ni(OH)_{2(s)} + 2\ H^+{}_{(aq)} \longrightarrow Ni^{2+}{}_{(aq)}\ (green) + 2\ H_2O$$

$$Ni^{2+}{}_{(aq)} + 6\ NH_{3(aq)} \longrightarrow Ni(NH_3)_6^{2+}{}_{(aq)}\ (green)$$

Step 3. Add 6 M HCl drop-wise to the precipitate from Step 2 until it has completely dissolved. Add 2 ml of water and 2 drops of 0.5 M KSCN. The presence of Fe^{3+} is confirmed by the formation of a blood-red solution of soluble $FeSCN^{2+}$.

$$Fe(OH)_{3(s)} + 3\,H^+_{(aq)} \longrightarrow Fe^{3+}_{(aq)} + 3\,H_2O$$

$$Fe^{3+}_{(aq)} + SCN^-_{(aq)} \longrightarrow FeSCN^{2+}_{(aq)}\text{ (blood-red)}$$

Step 4. Add 5 drops of dimethylglyoxime solution (abbreviated as H_2DMG) to the supernatant from Step 2. The formation of a bright red precipitate of nickel dimethylglyoxime, $Ni(HDMG)_2$, confirms the presence of Ni^{2+}.

$$Ni(NH_3)_6^{2+}{}_{(aq)} + 2\,H_2DMG \longrightarrow Ni(HDMG)_{2(s)}\text{ (red)} + 2\,NH_4^+ + 4\,NH_3$$

Step 5. Add 15 M HNO_3 drop-wise to the supernatant from Step 1 until it is acidic to litmus. Now add 15 M NH_3 drop-wise until the solution is basic to litmus. This will precipitate Al^{3+} as $Al(OH)_3$. Aluminum hydroxide is a white gelatinous substance but it may appear yellow due to the presence of unremoved CrO_4^{2-} ions on the surface of the precipitate. The $Al(OH)_3$ precipitate may appear as a suspended ring of gelatinous solid resembling a "smoke ring" in the center of the test tube. Due to the fragile nature of the precipitate, it is best to transfer it with a capillary or disposable pipet to another test tube for testing in Step 6.

$$Al(OH)_4^-{}_{(aq)} + 4\,H^+_{(aq)} \longrightarrow Al^{3+}_{(aq)} + 4\,H_2O$$

$$Al^{3+}_{(aq)} + 3\,OH^-_{(aq)} \longrightarrow Al(OH)_{3(s)}\text{ (white gelatinous)}$$

Step 6. To the precipitate from Step 5, add 6 M HNO_3 drop-wise until the precipitate dissolves. To this solution, add 2 drops of aluminon dye reagent (aurin tricarboxylic acid) and then make the solution basic to litmus by adding 6 M NH_3. The formation of a "red lake" confirms the presence of aluminum in the sample. The aluminon dye colors the gelatinous $Al(OH)_3$ red and it appears as pink specks in the mixture.

The aluminon confirmation of Al^{3+} is often difficult to see. Therefore, an additional test using a UV lamp is suggested. Place a drop of the solution containing the "red lake" on a piece of filter paper. To this spot add a drop of freshly prepared Morin solution and a drop of 1 M HCl. Observe the spot under UV light for aqua-green fluorescence, which confirms the presence of Al^{3+}.

$$Al(OH)_{3(s)} + \text{aluminon reagent} \longrightarrow Al(OH)_3\text{ aluminon complex}_{(s)}\text{ (red lake)}$$

Step 7. Add 5 drops of 1 M $BaCl_2$ to the supernatant from Step 5. The presence of CrO_4^{2-} is confirmed by the formation of a light yellow precipitate. Be patient for the $BaCrO_4$ precipitate to form. To hasten crystallization, scratch the inside of the test tube with a glass stirring rod.

$$CrO_4^{2-}{}_{(aq)} + Ba^{2+}_{(aq)} \longrightarrow BaCrO_{4(s)}\text{ (light yellow)}$$

Safety Precautions

Be cautious when using ammonium hydroxide, nitric acid, and sodium hydroxide as these are caustic acids and bases and may cause severe burns of skin and holes in clothes. Wash your hands thoroughly after this experiment to remove any traces of chromium, aluminum, nickel, and iron ions.

Waste Management

Pour all waste from this experiment into the "Metals waste" container in the fume hood.

Qualitative Analysis: Group III
Prelab Questions

Name _____ **Section** _____

1. Name the ions that remain in the supernatant in Step 1 of the Group III analysis.

2. Which metal reacts to give a blood-red color to the solution during its confirmatory step?

3. Which precipitate looks like a gelatinous "smoke ring" in the test tube?

Qualitative Analysis: Group III
Data Sheet

Name _____ **Section** _____

Name of Lab Partner(s) _____

Group III Unknown Number _____

Step	Known	Unknown
1. Observations		
2. Observations		
3. Observations		
4. Observations		
5. Observations		
6. Observations		
7. Observations		

Ions Identified in Unknown _____ _____ _____ _____

Qualitative Analysis: Group III
Review Questions

Name _____ **Section** _____

1. Write balanced **net ionic equations** for all of the precipitation reactions that you observed with your unknown. You should have one equation for every precipitate noted in the "unknown" column of your data sheet.

2. (a) After the addition of $NaOH/H_2O_2$ in the initial step, which Group III ions remain soluble in the supernatant? (b) How are these ions separated for analysis? (c) Write net ionic equations for the confirmation reaction for each of these ions.

3. A student treated an unknown with NaOH and H_2O_2 and discovered that a precipitate and <u>colorless</u> supernatant resulted. (a) What ions are possibly present? (b) If the treatment of the unknown had resulted in <u>only</u> a colorless supernatant, what ion could be present? Explain.

4. In the Group III qualitative analysis scheme, acids are often added followed by the addition of a base. What is the reason for this procedure?

Qualitative Analysis: Group III
Supplemental Questions

Name _____ Section _____

Significant Figures and Density

Introduction

In this laboratory exercise, you will be introduced to the concepts of measurement and units. You will gain an understanding of how to collect data and how to interpret uncertainties in that data in order to provide the best experimental value. Knowledge of some basic physical properties of substances will also be gained.

In reporting a measurement, the researcher always reports a number corresponding to the best value for that measurement. It is very important to include **the units of the value** which describe how the value was determined. Whenever an experiment is performed, there is always some uncertainty involved in the final result obtained. Therefore, the researcher must consider the type of tool used in making the measurement and how well the measurement was taken in order to report the best value. Reporting the final result for a particular measurement takes into consideration the reliability of the data collected.

A scientist will generally repeat an experiment several times. These repetitions are called **trials**. Usually, the values obtained from a set of trials will encompass a range of values. Therefore, it is necessary to determine the best value to report for the final result of the experiment. The final result may be an average of the values obtained from a set of trials or some other statistic that represents the best value. The researcher can determine **precision**, or reproduciblity of measurements, in terms of how wide the range of values is in the set.

Consider the dartboard in Figure 1(a). For three replicate dart throws, the difference between dart locations is very small. Thus, the person throwing the darts was very precise as there is little deviation in location of the darts. However, the dartboard in Figure 1(b) illustrates a person who is not only precise in their throwing, but also **accurate**, as the darts hit the bulls-eye. Accuracy is determined by how close a measurement is to the accepted standard or true value.

It is important to note that precision can be achieved without accuracy and conversely, an accurate value may be obtained without having good reproducibility, or precision. Scientists generally interpret an experiment that has both good accuracy and precision as one that was carried out very well.

The fact that any measured value is limited by the reliability of the measurement is conveyed by the amount of digits in the final reported value. Generally, the very last digit reported is the least certain digit according to the precision of the analysis. To correctly assign a value which takes into consideration the reliability of the measurement, the experimenter should always use the rules of the **significant figures convention**. *The significant figures convention states that the value reported for a measurement is rounded off so that it contains only the digits known with certainty plus one uncertain one which is the last digit.*

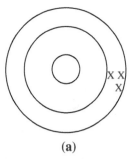 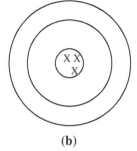

(a)   **(b)**

Good Precision but Poor Accuracy   **Good Precision and Good Accuracy**

FIGURE 1 — *Accuracy and Precision Dartboard Illustrations*

Consider, for example, a measurement of distance made with a wooden ruler and reported as 3.28 cm. This implies that the distance lies between 3.2 cm and 3.3 cm, and that the distance is considerably closer to 3.3 cm. Reporting the distance as 3.2846 cm would make no sense because the tool used to make the measurement is not likely to be that accurate. The recorded measurement of 3.28 cm has only one somewhat doubtful or uncertain digit (the "8") and has been rounded off to remove all non-significant digits.

When performing mathematical computations, the final calculated value which is reported can only be as good, or as certain, as the least certain value in the calculation. Consider the calculation performed on a pocket calculator in the following example:

$$\text{Density} = \frac{\text{mass}}{\text{volume}} \qquad D = \frac{15.75 \text{ g}}{12.4 \text{ mL}} = 1.27016129 \text{ g/mL}$$

The calculator gives many more digits, suggesting a much greater precision than justified by the numbers in the calculation! The significant figure rules, when applied to this calculation, yield a density value as 1.27 g/mL. Now we have a more reasonable answer which is within the limitations of our measurements.

Significant Figure Rules

A value that has more significant digits is one that gives more information about a particular measurement and its precision. Note that defined quantities (one dozen = 12; 2 cups = 1 pint; 100 cm = 1 m) and counted numbers (2 apples; 4 tires) are not considered in our significant figure rules as they are exactly known.

To determine the significant figures in a reported value, read the number from left to right and count all digits, starting with the first digit that is not zero. In any measurement that is properly reported, all nonzero digits are significant. Zeros, however, can be used either as part of the measured value or merely to locate the decimal point. Thus, zeros may or may not be significant, depending on how they appear in the number. The following guidelines describe the different situations involving zeroes (excerpt taken from *Chemistry: The Central Science*, 11th edition by Brown, LeMay, Bursten, and Murphy):

1. Zeroes between nonzero digits are always significant. For example: 1005 kg contains four significant figures; 1.03 contains three significant figures

2. Zeroes at the beginning of a number are never significant; they merely indicate the position of the decimal point. For example: 0.02 g (one significant figure); 0.0026 (two significant figures).

3. Zeros at the end of a number are significant if the number contains a decimal point. For example: 0.0200g (three significant figures); 3.0 cm (two significant figures).

Scientific notation may be used to convey a small amount of significant digits when a value is large. For example, it is acceptable to report a calculated value of 180,000 as 1.80×10^5 if you are limited to only 3 significant digits due to the calculation.

The number of significant figures (abbreviated Sig.Fig. or just S.F.) in the following examples are given in parentheses.

123.45	(5 S.F.)
8.4381	(5 S.F.)
0.00468	(3 S.F.)
3033	(4 S.F.)
20,000	(1 S.F.)
35,000.0	(6 S.F.)
0.000430	(3 S.F.)
0.08000	(4 S.F.)
100,004	(6 S.F.)

Significant Figures in Calculations

1. When **adding or subtracting**, the number of decimal places in the answer should be equal to the number of decimal places in the value with the fewest decimal places.

   Example:
   0.00453 g (5 decimal places)
   46.208 g (3 decimal places)
   1.9 g (1 decimal place)
   _____
   48.11253 g

   We are limited to just one decimal place, so the answer should be reported as 48.1 g.

2. In **multiplication or division**, the number of significant figures in the answer should be the same as the value with the <u>fewest significant figures</u>.

   For example $\dfrac{1.2758 \text{ g}}{2.1 \text{ mL}}$ = 0.6065238 g/mL

   We are limited by the 2 S.F. in the value 2.1 mL, so the answer should be reported as: 0.61 g/mL.

3. When a **number is rounded off** (to decrease the number of significant figures reported) the last digit is increased by 1 only if the following digit is 5 or greater.

   Example: 25.765 cm rounded off to 4 S.F. becomes 25.77 cm.
   3.82 mL rounded off to 2 S.F. becomes 3.8 mL.

*It is important to note that the final answer cannot have more significant digits than the least precise piece of information that went into the calculation.

Significant Figures in Measurements

Now, we must consider how to report numerical values from measurements made in the laboratory. Always **record ALL digits** an instrument provides in order to achieve the best value for your measurement. Do not add 0's or any other number to that value in order to have it "match" the number of decimal places in other measurements. _Assume the value shown on an instrument's display is the best, or most precise, for that instrument_. Do not attempt to weigh "exactly" 10.00 grams as the procedure for an experiment may instruct. Rather, weigh approximately 10 grams of a substance as best as you can, recording all the digits displayed by the balance.

For measurements which are obtained visually, such as reading centimeters off a ruler or milliliters from a graduated cylinder, it is often necessary to estimate between markings (interpolation). The interpolated digit would be the least certain of the significant digits in your value, but it is still a valid significant digit. In observing quantities of a solution in a graduated cylinder for example, you can interpolate between graduations on the cylinder. Consider the example in Figure 2. Reading a volume (in milliliters) from a 50 mL graduated cylinder in which the graduations

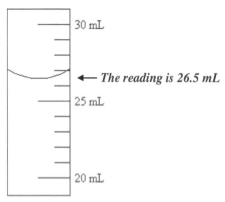

FIGURE 2 — *Volume Measurement using a 50 mL Graduated Cylinder*

are every 1 mL will allow you to estimate volume to the nearest 0.1 mL. Thus, the uncertainty in this reading will be in tenths of a mL.

Procedure

Part A: Significant Figures in Volume Measurements

You will measure the volume of water several different ways and record the number of significant digits in each value next to the line "# of Sig Figs". In your observations, note how water forms a concave curved surface in a narrow container. This **meniscus** is due to the forces of attraction and repulsion between the liquid and the sides of the container. You can place a piece of white paper behind the container to better read the volume if you desire. Record the volume of the liquid at the bottom of the meniscus. It is necessary to start with <u>clean and dry</u> glassware when making volume measurements to ensure precise and accurate readings.

a) Pour approximately eight mL of water into your 50 mL graduated cylinder. (Note that you are estimating the first decimal place). Record the volume on the data sheet.

b) Pour approximately eight mL of water into your 10 mL graduated cylinder. Read and record this volume on the data sheet to the appropriate number of significant figures.

c) Place some water (greater than 10 mL) in a 50 mL beaker. Obtain a 10 mL Mohr pipet (graduated pipet) and after practicing the proper use, withdraw approximately 8 mL of water from the beaker into the pipet. Record the volume to the nearest 0.01 mL on your data sheet.

d) Dispense the 8 mL of water from the pipet into a preweighed 50 mL beaker. (Be sure that the mass of the beaker is recorded on your data sheet). Reweigh the beaker with the water in it, determine the mass of the water, and record this mass on the data sheet. Knowing the density of water is 0.997044 g/mL at room temperature (25 °C), calculate the volume of water from the mass and density.

Part B: Density

Density is a common physical property which is often used by scientists to identify substances. It is an **intensive physical property**, meaning that its value is independent of the amount of material being measured. Intensive physical properties are so important to the identification of a substance that much time has been devoted to collecting these data and compiling them into the *Handbook of Chemistry and Physics* (The Chemical Rubber Co) and the *Handbook of Chemistry* (N.A. Lange).

Density is defined as the mass of an object or substance divided by its volume ($d = m/v$). In the chemistry lab, we typically use units of g/mL or g/cm³ to express the density of substances. While there is no physical difference between a g/mL and a g/cm³, it is common to report the densities of liquids in units of g/mL and the densities of solids in units of g/cm³. The densities of gases are usually expressed in units of g/L since gases have very small masses for a given volume.

The mass of a sample can be determined using a balance. If the substance is a liquid, volume can be determined by observing the level of the liquid in a piece of graduated glassware. If the substance is a solid, the volume can be determined by observing the volume of liquid displaced by the substance. If the solid has a regular shape, its volume can be calculated from the following relationships:

<u>rectangle</u>	volume = length × width × height
<u>cube</u>	volume = (length)³
<u>cylinder</u>	volume = π × (r²) × height

In this experiment you will measure the mass and volume of a metal in order to determine its density. From the density values you will identify the metal as one of the following:

Al (d = 2.702 g/mL), Ni (d = 8.90 g/mL),
Cu (d = 8.92 g/mL), Pb (d = 11.35 g/mL)

Select an unknown metal. You are to determine the metal's density by using three different samples of the same unknown metal. For the first trial, determine the mass of approximately 20 g of the metal to the nearest 0.01 g (or to the nearest milligram depending on the balance used). Record the mass of the metal on the data sheet. Pour water (at least 20 mL) into a 50 mL graduated cylinder. Record the volume of water. Add the metal to the water and record the new volume reading. Calculate the difference in volume on the data sheet. This difference represents the volume of the metal. From the mass of the metal and the volume it displaced, calculate the density of the metal. Be certain the metal is **DRY** before returning it to the appropriate bottle.

Repeat this procedure 2 more times using two new samples of the same unknown metal. Vary the mass of metal for the second and third trial (e.g. approximately 25 g; 30 g). Use your dried 50 mL graduated cylinder and refill it with approximately 20 mL of water. As you are only interested in the difference in volume due to the displacement by the metal, it does not matter how much liquid you use provided you record the precise amount. Record the data and calculate the density for each trial.

Average the density measurements to determine the density of your metal and identify it. Return your data sheet and all of your question sheets to your instructor.

Safety Precautions

Wash your hands with soap and water thoroughly after you have completed the experiment since you may have handled lead metal.

Waste Management

Collect all your unknown metal sample from your lab bench and from the electronic balance area. Carefully dry the metal sample without losing any in the sink or on the floor. Please return metals to their original container on the reagent bench.

Significant Figures and Density
Prelab Questions

Name _____ Section _____

1. Express the following value for the speed of light to 7 significant digits: 29,979,300,000 cm/sec

2. Circle the following physical properties which depend on the quantity of material (extensive physical properties):

    boiling point
    density
    mass
    color
    volume
    melting point

3. Record the volume (in mL) contained in each piece of glassware pictured below. Carefully consider the scale in each piece when recording decimal places.

 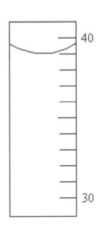 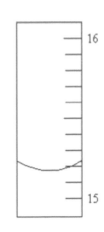

Significant Figures and Density Data Sheet

Name _____ Section _____

Name of Lab Partner(s) _____

Part A: Significant Figures in Measurements

Volume of water in 50 mL graduated cylinder: _____ # of Sig Figs: _____

Volume of water in 10 mL graduated cylinder: _____ # of Sig Figs: _____

Volume of water in 10 ml pipet: _____ # of Sig Figs: _____

Mass of Water & Beaker: _____ # of Sig Figs: _____

Mass of Beaker: _____ # of Sig Figs: _____

Mass of Water: _____ # of Sig Figs: _____

Volume of Water in Beaker (mass/density = volume): _____ # of Sig Figs: _____

Part B: Density Unknown Metal # _____

	Trial 1	Trial 2	Trial 3
Mass of metal (g)			
Grad. Cylinder reading with H_2O and metal (mL)			
Grad. Cylinder reading with H_2O only (mL)			
Metal volume (mL) (change in volume)			
Calculated Density (g/mL)			

Average Density: _____   Identity of Metal: _____

1. One way to determine the **precision** of your data set is to calculate what is called the average absolute deviation. To do this, subtract each calculated density value for each trial from the average density value for the unknown metal. This is called the deviation. Make each deviation positive (absolute value) and then calculate the average deviation by dividing the sum of the deviations by three. What is the precision of your unknown metal's average density?_____

2. Was your average density value **accurate** to the known metal values within ± 1.0 g/ml?

Significant Figures and Density
Review Questions

Name _____ Section _____

1. a) A cube of titanium metal measures 1.32 cm on each side. It has a mass of 10.35 grams. What is the density of this metal?

   b) What is the density of a cube of titanium that has a volume of 5.43 cm^3?

2. Indicate the number of significant figures in each of the following measured quantities:

   a) 1.09×10^{-3} km
   b) 0.0234 m^2
   c) 7,300 cm
   d) 8.9326 g
   e) 20.0 mL

3. A bar of gold weighs 5.95 lb. If the density of gold is 19.31 g/mL, what is the volume of the bar? Report the answer in the correct number of significant figures. (454 g =1 lb).

4. If I drink a can of Pepsi (355.0 mL), what mass of drink am I consuming if Pepsi has a density of 1.21 g/mL? Show all work. Report the answer to the correct number of significant figures.

5. Two students determine the percentage of lead in a sample as a laboratory exercise. The true percentage is known to be 23.65. The students' results for three determinations are shown below:

Student 1: 23.66, 23.78, 23.47

Student 2: 23.42, 23.38, 23.40

a) Calculate the average for each set of data. Explain why one set is more accurate than the other.

b) One way to determine the precision of each data set is to calculate what is called the average absolute deviation. To do this, subtract each data point in a set from the average value for the set. This is called the deviation. Make each deviation positive (absolute value) and then calculate the average deviation for each data set. Which data set has the greatest precision?

Significant Figures and Density
Supplemental Questions

Name _____ **Section** _____

Solubility Product: K_{sp} of PbI_2

Introduction

This is the second of two experiments described in your laboratory manual that involves measuring the value of an equilibrium constant. The first of these experiments, *Equilibrium Constant, Keq*, contains an introduction to the theory and terminology of chemical equilibria and should be read before conducting this experiment.

In today's exercise, you will measure the value of the equilibrium constant for the dissolution of solid PbI_2 in a saturated solution of PbI_2. This reaction is an example of a general class of equilibrium processes known as **solubility equilibria**. The equilibrium constants for solubility equilibria are given the special notation K_{sp} (**solubility product constant**). The equilibrium equation for this process and the equilibrium expression are shown below.

$$PbI_{2(s)} \rightleftharpoons Pb^{2+}_{(aq)} + 2\,I^-_{(aq)}$$

$$K_{sp} = [Pb^{2+}]\,[I^-]^2$$

Notice that $[PbI_{2(s)}]$ does not appear in the denominator of the equilibrium expression. The reason for this is that the concentrations of pure solids, liquids, and solvents remain constant in any chemical system, so their values do not appear in equilibrium expressions.

The equilibrium concentration of iodide ion (I^-) will be determined spectrophotometrically *after* converting the colorless iodide ion into iodine (I_2, yellow-brown in aqueous solution). Iodide ions will be oxidized to iodine molecules using KNO_2.

$$2\,NO_2^-{}_{(aq)} + 2\,I^-{}_{(aq)} + 4\,H^+{}_{(aq)} \rightleftharpoons 2\,NO_{(g)} + 2\,H_2O_{(l)} + I_{2(aq)}$$
$$\text{\textit{colorless}} \qquad\qquad\qquad \text{\textit{yellow-brown}}$$

The stoichiometric factor for Pb^{2+} and I^- in this reaction is 1:2, therefore, the equilibrium concentration of Pb^{2+} is found by dividing the equilibrium concentration of I^- by 2.

The absorbance of the solutions in this experiment will be measured at a wavelength between 470 and 530 nm using the MicroLab spectrophotometer. A detailed introduction to the theory and use of the MicroLab spectrophotometer is contained in **Appendix D**.

Procedure

All of your absorbance readings should be made using glass **cuvettes** designed to fit in the MicroLab spectrophotometer. A cuvette is small glass container of uniform thickness and clarity that is specifically designed for making spectrophotometric measurements. It is important that you do not write on the cuvette lids.

Part A: Creating a Calibration Curve

Place four clean medium test tubes in your test tube rack. Label them 1, 2, 3, and blank. The amounts of reagents to add to each test tube are summarized in Table 1. Use a 10 mL graduated pipet to deliver 3.00 mL of 0.030 M KI to the first tube, 2.00 mL to the second, 1.00 mL to the third, and none to the blank. Rinse the pipet with deionized water and then deliver 1.00 mL of deionized water to the first test tube, 2.00 mL to the second test tube, 3.00 mL to the third, and 4.00 mL to the blank. Rinse the pipet and then deliver 8.00 mL of 0.020 M KNO_2 (not to be confused with KNO_3) into each of the four test tubes. Add 2 *__drops__* *(not mL)* of 6 M HCl to each tube and stir well. All of the solutions except the blank should have a yellow-brown color.

Pour each of your four solutions into clean, dry cuvettes until they are about 3/4-full. The reaction between NO_2^- and I^- produces $NO_{(g)}$ as one of the products. Use a glass stirring rod to dislodge any gas bubbles sticking to the sides of the cuvette in order to avoid erroneous absorbance readings.

Table 1 *Volumes of Reagents and Concentration of I^-*

Test Tube	KI (mL)	DI water (mL)	KNO_2 (mL)	HCl (*__Drops__*)	Concentration of I^- (M)
1	3	1	8	2	0.0075 M
2	2	2	8	2	0.0050 M
3	1	3	8	2	0.0025 M
blank	0	4	8	2	0

Measuring Absorbance:

Set up your laboratory computer and the MicroLab data acquisition system. Start the MicroLab software and select Spectrophotometer Experiment.

1. Place the blank in the spectrophotometer and press the "Read Blank" button.

2. Click on "Absorbance" at the top of the screen. This will switch the spectrophotometer readings from % transmittance to absorbance.

3. Remove the blank and place Standard 1 (0.0075 M I^-) in the spectrophotometer. Press the "Add" button. In the dialog box that appears, enter "Std 1" for the Sample ID and enter the actual concentration of the standard (i.e. 0.0075) in the Concentration box. Press the "OK" button.

4. Click on the wavelength bar that represents the optimum response between wavelengths 470 and 530 nm. On your data sheet provide 2 reasons for your wavelength choice.

5. Repeat step (3) for the remaining standards labeling them "Std 2" and "Std 3". Record the absorbance values for each standard on your data sheet.

6. Click on the tab labeled "3 Curve" and select "linear" for the curve fit. The computer will automatically generate your calibration curve. Record the equation for your calibration curve on your data sheet along with the correlation coefficient reported by the MicroLab software. You will use this calibration curve to determine the equilibrium concentration of I^- in Part C.

7. Keep your standards available until the entire experiment is complete in the event you need to restore your curve on the computer. When entire experiment is finished, wash the cuvettes and lids well with soap and water, and rinse with DI water two times before returning to the front reagent bench. Do **not** use wire brushes to clean the cuvettes.

Part B: Preparation of Sample

Use your 10 mL graduated cylinder to measure 10 mL of 0.012 M $Pb(NO_3)_2$ (dissolved in KNO_3) into a medium test tube. Rinse the graduated cylinder and use it to add 10 mL of 0.030 M KI (dissolved in KNO_3) to the same test tube[1]. Stopper the test tube with a cork and shake it for 15 seconds. Set up a suction filtration apparatus like the one shown in Figure 1 and place a piece of filter paper in the Büchner Funnel. Turn on the suction and moisten the filter paper with deionized water.

Carefully pour the contents of the medium test tube into the Büchner funnel. Wash the yellow solid with one 15 mL portion of deionized water. Using your spatula, remove the filter paper containing the PbI_2 from the Büchner funnel and place it in a medium test tube. (The filter paper must be curled into a cylinder to fit into the test tube). Add 10 mL of deionized water to the test tube, stopper, and shake for 10–15 seconds. Remove the filter paper from the test tube (before it disintegrates), restopper it, and shake the solution for 20 seconds every minute for the next 10 minutes.

Part C: Determination of [I⁻] in a Saturated Solution of PbI_2

Allow the contents of the test tube from Part B to settle. Fold a piece of 11 cm filter paper according to your lab instructor's instructions and place it in a funnel. Decant the liquid from the test tube into the funnel. Let the funnel drain into a clean dry, medium test tube. If any solid PbI_2 comes through with the filtrate, the solution will have to be filtered again. Use a 10 mL graduated pipet to transfer 4.00 mL of the filtrate to another clean, dry, medium test tube.

Add 8.00 mL of 0.020 M KNO_2 and 2 ***drops*** of 6 M HCl and stir well. Pour some of this solution into a cuvette until it is about 3/4-full. Place this cuvette in the Microlab spectrophotometer. Click on the tab labeled "4 Read". Press the "Add" button and enter "Sample 1" for the Sample ID. Press the "OK" button. The computer will now display the concentration of I⁻ in this sample. Record the absorbance and concentration values for I⁻ in Sample 1 on your data sheet.

Multiply the concentration value for I⁻ in sample 1 by a factor of 3 to obtain the equilibrium concentration of I⁻ in the saturated solution of PbI_2. This is necessary because the original saturated solution was diluted by placing 4 mL in a total of 12 mL of solution (a 1 in 3 dilution). Record this value on your data sheet as the concentration of I⁻ in

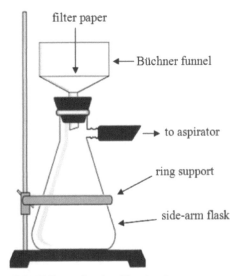

FIGURE 1 — *Suction Filtration Apparatus*

---

[1]The $Pb(NO_3)_2$ and KI are dissolved in KNO_3 to aid in the formation of large crystals of PbI_2.

saturated PbI_2. Calculate the concentration of Pb^{+2} using information about the stochiometry of the dissolution of PbI_2 (the ratio of Pb^{+2} to I^- is 1:2). Now you have all the information you need to calculate the K_{sp} value of $PbI_{2\,(s)}$.

Clean the Büchner funnel and side arm flask well with soapy water and rinses with DI water and return them to their designated location on the shelf. Wash all the cuvettes and lids well with soap and water, and rinse with DI water two times before returning to the front reagent bench. Do **not** use wire brushes to clean the cuvettes.

Safety Precautions

Compounds containing lead are poisonous. Be sure to wash your hands and under your fingernails before leaving the laboratory. Potassium nitrite (KNO_2) is a strong oxidizing agent. Be careful not to get this solution on your skin or clothes.

Waste Management

Place all liquid wastes from Parts B and C in the container labeled "Metals waste" because they contain lead. Solid wastes (filter papers) from Parts B and C may be wrapped in a paper towel and thrown away in the trash can.

Wastes from Part A can be flushed down the sink with plenty of water.

Solubility Product: K_{sp} of PbI_2
Prelab Questions

Name _____ **Section** _____

1. Write the Beer-Lambert Law and define its terms. What two variables from the Beer-Lambert Law will you use to construct your calibration curve? (Refer to **Appendix D** for a discussion of this law.)

2. Write the balanced chemical equation for the preparation of $PbI_{2\,(s)}$ in the experiment.

3. Discuss the distinction between the use of the 0.03 M KI solution (dissolved in KNO_3) and the 0.03 M KI solution (dissolved in DI water) in this experiment. Why are these solutions prepared differently?

4. What volume of 0.030 M KI solution is needed to prepare 6.0 mL of a 0.0075 M KI standard? (Hint: use the formula $M_1V_1 = M_2V_2$)

Solubility Product: K_{sp} of PbI_2
Data Sheet

Name _____ **Section** _____

Name of Lab Partner(s) _____

Part A

Wavelength (nm) _____

Reasons wavelength chosen: _____

Absorbance of Standard 1, 0.0075 M I⁻ _____

Absorbance of Standard 2, 0.0050 M I⁻ _____

Absorbance of Standard 3, 0.0025 M I⁻ _____

Calibration Curve Equation from Microlab Software _____

Correlation Coefficient _____

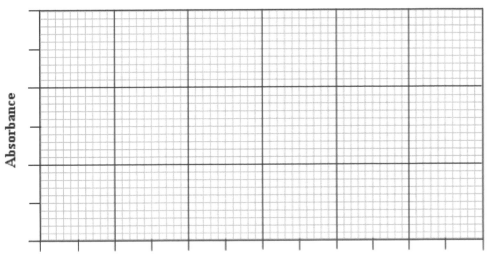

Part C

1. Absorbance of [I⁻] in Sample 1 _____

2. Concentration of [I⁻] in Sample 1 _____

3. Concentration of [I⁻] in saturated PbI_2 _____ (requires using dilution factor)

4. Concentration of [Pb^{2+}] in saturated PbI_2 _____

5. $K_{sp} = [Pb^{2+}][I^-]^2 =$ _____

Solubility Product: K_{sp} of PbI_2
Review Questions

Name _____ **Section** _____

1. Describe what your K_{sp} value means as related to this reaction. In your explanation, discuss the main concept of K_{sp} as if you were talking to someone unfamiliar with this experiment. (Hint: is the dissolution of $PbI_{2\,(s)}$ reactant favored or product favored? Explain what this implies about $PbI_{2\,(s)}$.)

2. Could the calibration curve prepared in this experiment be used for other iodide salts such as NaI or LiI? Could this calibration curve be used at other wavelengths ? Explain.

3. The concentration of X in a saturated solution of XY_3 is determined to be 0.001 M. What is the concentration of Y ? What is the K_{sp} of the salt ? Show your calculations.

4. Which of the following ionic compounds is the most soluble in H_2O at 25 °C? Which ionic compound is least soluble in water at 25 °C?

Salt	K_{sp}
$Fe(OH)_3$	6.3×10^{-38}
$PbBr_2$	6.3×10^{-6}
MnF_2	6.4×10^{-9}
Hg_2SO_4	6.8×10^{-9}

Solubility Product: K$_{sp}$ of PbI$_2$
Supplemental Questions

Name _____ **Section** _____

Spot Tests: Another Qualitative Analysis

Introduction

The previous two qualitative analysis experiments, *Qualitative Analysis: Groups I and III*, tested for the presence of cations in a solution utilizing an approach that systematically separated ions from each other based on their solubility differences. The goal in each of those experiments was to separate and isolate each cation and then conduct a confirmatory test for the ion. Since the confirmatory test was conducted on a solution that contained, at the most, only one ion, there was no need to worry about the presence of ions that might interfere with the confirmatory test. The object of today's experiment is to *test for specific ions in a mixture containing other ions* using a technique called **spot tests**. Since you will be conducting your tests on a mixture of ions, you will need to consider how the presence of other ions could interfere with the results of your test for a specific ion. Ions that can interfere with the tests for a specific ion are, not surprisingly, called **interfering ions**.

An example of a problem posed by interfering ions is found in the spot test for the chloride ion, Cl^-. This test involves adding Ag^+ to a solution that might contain chloride ions. If Cl^- is present, a white precipitate of $AgCl$ forms. However, other ions such as SCN^- also form white precipitates with Ag^+. Therefore, adding Ag^+ to a solution can confirm the presence of Cl^- only if SCN^- is *not* present. SCN^- is said to be an interfering ion in the Cl^- test. You will need to consider problems like these when analyzing the results of your spot tests.

Procedure

The ions you will test for in this experiment are: Cl^- (chloride), SCN^- (thiocyanate), $C_2H_3O_2^-$ (acetate), PO_4^{3-} (phosphate), CO_3^{2-} (carbonate), CrO_4^{2-} (chromate), SO_4^{2-} (sulfate), and NH_4^+ (ammonium). You will find bottles containing solutions of each of these ions, to be used as knowns, on the dispensing shelf.

Spot tests on knowns and unknowns should be performed simultaneously. Each spot test of the unknown is performed on a fresh 1 mL sample of the unknown. Your unknown will contain from one to four of the ions shown above. When conducting your tests, keep in mind that your unknown is diluted and perhaps colored by the presence of several ions. Therefore, positive test results might have slightly different appearances between the known and your unknown.

Some of the spot test reactions occur rather slowly. If no reaction is apparent after the addition of a reagent, place the test tube in a hot water bath for 1 to 3 minutes. Be sure to stir each solution after the addition of a test reagent.

The following tests should be performed in the order shown. They have been arranged so that possible interfering ions are identified first. *Be very careful not to contaminate your unknown with the knowns or with any of the test reagents.*

Test 1. Carbonate Ion

Place 5 drops of 1 M Na_2CO_3 in a micro test tube followed by 5 drops of 6 M HCl. The formation of bubbles indicates release of carbon dioxide gas. Bubbles are not quite as evident in the unknown solution because the unknown is diluted by other ions. Repeat the test using 5 drops of your unknown. If no reaction is observed, repeat the test, warming it in a hot water bath for 2 minutes.

$$CO_3^{2-} + 2\,H^+ \longrightarrow CO_{2\,(g)} + H_2O$$

Test 2. Thiocyanate Ion

In the *Qualitative Analysis: Group III* experiment you utilized a test which identified iron(III) by its reaction with thiocyanate ion, SCN^-, to form a blood-red complex. This is the basis of the thiocyanate spot test. Add 5 drops of 6 M HCl to 5 drops of 0.5 M KSCN on the spot plate. Add 1 drop of 0.1M $Fe(NO_3)_3$ to the spot plate. The solution will turn red if SCN^- is present. Repeat the test on your unknown.

$$SCN^- + Fe^{3+} \longrightarrow FeSCN^{2+} \text{ (red)}$$

Test 3. Sulfate Ion

Put 5 drops of 0.5 M Na_2SO_4 in a micro test tube. Add 5 drops of 6 M HCl to the test tube. Now add 1 drop of 1 M $BaCl_2$ to the test tube. A precipitate of white $BaSO_4$ indicates the presence of sulfate ion. Repeat the test on your unknown.

$$SO_4^{2-} + Ba^{2+} \longrightarrow BaSO_{4\,(s)} \text{ (white)}$$

Test 4. Phosphate Ion

Place 5 drops of 0.5 M Na_2HPO_4 (sodium monohydrogen phosphate) in a medium test tube. Add 1 mL of 6 M HNO_3. Now add 1 mL of 0.5 M $(NH_4)_2MoO_4$ (ammonium molybdate) to the test tube. The presence of phosphate ions is confirmed by the formation of a yellow precipitate of $(NH_4)_3PO_4 \cdot 12\,MoO_3$ (ammonium phosphomolybdate). This precipitate is often slow to form and heating in the hot water bath may be required to initiate precipitation. Repeat the test on your unknown.

$$PO_4^{3-} + 24\,H^+ + 12\,(NH_4)_2MoO_4 \longrightarrow (NH_4)_3PO_4 \bullet 12\,MoO_{3\,(s)} + 21\,NH_4^+ + 12\,H_2O$$

This test is the basis for the analysis of phosphate ion in water supplies. Phosphate ion is a nutrient for microorganisms and plants, and in high quantities is indicative of pollution.

Test 5. Chromate Ion

Solutions containing chromate ions are yellow when neutral or basic and orange when acidic. The orange color is due to the formation of the dichromate ion $(Cr_2O_7^{2-})$. Add 5 drops of 0.5 M K_2CrO_4 to a spot plate. Add 5 drops of 3% H_2O_2. Now add 10 drops of 6 M HNO_3 to the plate. The formation of a blue color, which appears but rapidly disappears, indicates the presence of the chromate ion. Repeat this test on your unknown.

$$2\,CrO_4^{2-} + 2\,H^+ \longrightarrow Cr_2O_7^{2-}{}_{(aq)} + H_2O$$

$$Cr_2O_7^{2-} + 2\,H^+ + 4\,H_2O_2 \longrightarrow 2\,CrO_5 \text{ (blue)} + 5\,H_2O$$

Test 6. Chloride Ion

Place 5 drops of 0.5 M NaCl in a micro test tube. Add 5 drops of 6 M HNO_3. Now add 2 drops of 0.1 M $AgNO_3$ to the test tube. The presence of chloride ion is indicated by the formation of a white precipitate of AgCl. If Test 2 indicated that your unknown contained SCN^-, then the SCN^- must be removed before you test your unknown for Cl^- because SCN^- also forms a white precipitate with Ag^+.

The following procedure is necessary only if your unknown contains SCN^-. Place 1 mL of your unknown in a 50 mL beaker. Add 2 mL of 6 HNO_3 to the beaker and boil the solution gently until about 1 mL of the solution remains. This will destroy any SCN^- that is present. Now add 1 mL of HNO_3 and 2 drops of $AgNO_3$ to the beaker. The formation of a white precipitate indicates the presence of Cl^-.

$$Cl^- + Ag^+ \longrightarrow AgCl_{(s)} \text{ (white)}$$

Test 7. Acetate Ion

Place 5 drops of 1 M $NaC_2H_3O_2$ (sodium acetate) in a micro test tube. (The acetate ion is often abbreviated as OAc^-). Add 5 drops of 3 M H_2SO_4 to the test tube. The presence of acetate ion is indicated by the smell of acetic acid (a vinegar odor) coming from the plate. If no odor is immediately present, warming the test tube in a hot water bath for 20 seconds and check the odor again. Repeat the test with your unknown.

$$OAc^- + H^+ \longrightarrow HOAc \text{ (acetic acid)}$$

Test 8. Ammonium Ion

This test is similar to Test 7 in that it relies on your sense of smell to detect the presence of a substance. Add 5 drops of 0.5 M NH_4Cl to a micro test tube. Add 5 drops of 6 M NaOH to the test tube and check for the presence of an ammonia odor. If no odor is immediately apparent, warm the tube for about 20 seconds and check again. You may test for ammonia fumes emerging from the test tube by holding a piece of red litmus paper over the tube while the solution is being warmed. Do not allow the litmus paper to touch the test tube. A change in litmus color from red to blue indicates the presence of ammonia, which is a weak base.

$$NH_4^+ + OH^- \longrightarrow H_2O + NH_3 \text{ (ammonia)}$$

As an additional test for NH_4^+, place a drop of the unknown on a spot plate and add a drop of Nessler's reagent. A reddish-brown precipitate indicates the presence of the NH_4^+ ion. Repeat the test on your unknown.

Interfering Ions

The results you obtain with your unknown will not always match the results with the knowns due to the presence of interfering ions. If you suspect that an ion is interfering with a particular test, make up a known that contains the ions you think you have in your unknown and repeat the spot tests on this solution.

The tests that are likely to suffer the most form interference problems are Test 3 and Test 6. In Test 3, Ba^{2+} is added to a solution to test for the presence of SO_4^{2-}. However, Ba^{2+} ions also form precipitates with CrO_4^{2-} and PO_4^{3-} that could also be present in your unknown. The colors of these precipitates are: $BaCrO_4$ (pale yellow), $BaCO_3$ (white), and $Ba_3(PO_4)_2$ (white).

In Test 6, Ag^+ is added to test for the presence of Cl^- ions. Ag^+ also reacts with others ions that could be present in your unknown to form the following insoluble compounds: Ag_3PO_4 (yellow), Ag_2SO_4 (slightly soluble, white), Ag_2CO_3 (pale yellow), Ag_2CrO_4 (red brown), and AgSCN (white). If Cl^- is present with any of these other anions, the AgCl can be dissolved with 10 drops of 6 M NH_3. Acidify the solution with 6 M HNO_3 and look for the presence of white AgCl.

Record all of your observations on your data sheet. You will write net ionic equations for all positive reactions in the Questions section.

Safety Precautions

Do not allow acids and bases to come in contact with your skin. Avoid skin contact with silver nitrate as it will leave a dark stain for several days.

Waste Management

Pour the unknown and known wastes from Test 2 and 4 of this experiment into the "Metals waste" container in the fume hood. Pour the unknown and known wastes from Test 5 into the "Chromates waste" container in the fume hood. Wastes from all other tests may be flushed down the sink with plenty of tap water.

Spot Tests: Another Qualitative Analysis Prelab Questions

Name _____ **Section** _____

1. What is the net ionic equation for the reaction that takes place when a solution containing PO_4^{3-} is treated with $AgNO_3$?

2. In the confirmatory test for SO_4^{2-}, the solution to be tested is treated with Ba^{2+}. Now, suppose that carbonate ion is also present and forms a precipitate. What is this precipitate? Write a net ionic equation for this reaction that interferes with the sulfate test.

Spot Tests: Another Qualitative Analysis Data Sheet

Name _____ **Section** _____

Name of Lab Partner(s) _____

Unknown Number _____

Ion	Known	Unknown
CO_3^{2-}		
SCN^-		
SO_4^{2-}		
PO_4^{3-}		
CrO_4^{2-}		
Cl^-		
$C_2H_3O_2^-$		
NH_4^+		

Ions Identified in Unknown: _____ _____ _____ _____

Spot Tests: Another Qualitative Analysis Review Questions

Name _____ **Section** _____

1. Write balanced **net ionic equations** for all of the **positive tests** that you observed with your unknown. Indicate if any precipitates were formed. You should have a net ionic equation for every positive test noted in the unknown column of your data sheet.

2. Two of the spot tests depend on the sense of smell in testing for ions. Write net ionic equations for these tests.

3. Two of the spot tests depend upon the formation of a white precipitate to confirm the presence of ions. One is a sulfate and one is a chloride. Write net ionic equations for these reactions.

4. A student added $BaCl_2$ to their unknown. A white precipitate formed. Can the student confirm that SO_4^{2-} is present? Explain. (Hint: see the section on interfering ions.)

Spot Tests: Another Qualitative Analysis Supplemental Questions

Name _____ Section _____

Stoichiometry: Loss of CO$_2$

Introduction

Chemical reactions follow the **Law of Conservation of Matter**, which states that matter is neither gained nor lost during a chemical reaction. A more modern interpretation of this concept is summarized in the **Law of Conservation of Atoms**, which states that atoms are neither created nor destroyed during a chemical reaction. In other words, the same number and type of atoms that were present in the reactants are also present in the products of a chemical reaction. A balanced chemical equation is a statement of the Law of Conservation of Atoms. The study of the quantitative relationships among reactants and products in chemical reactions is called **stoichiometry**.

Consider the following balanced chemical equation that represents the reaction between two ionic compounds, sodium iodide and lead(II) nitrate. The equation can be read as two moles of NaI react with one mole of Pb(NO$_3$)$_2$ to produce one mole of PbI$_2$ and two moles of NaNO$_3$. The coefficients in the equation tell us the relative amounts of each reactant that must react in order for the Law of Conservation of Atoms to be obeyed.

$$2\ NaI_{(aq)} + Pb(NO_3)_{2\ (aq)} \longrightarrow PbI_{2\ (s)} + 2\ NaNO_{3\ (aq)}$$

For example, if we know that 0.25 moles of Pb(NO$_3$)$_2$ have reacted, we know that it took 0.50 moles of NaI to react with it, because the ratio of NaI to Pb(NO$_3$)$_2$ in the balanced equation is 2:1. The ratio of the moles of one substance to the moles of another in a balanced chemical equation is called a **stoichiometric factor**. The balanced chemical equation shown above contains many stoichiometric factors. Here are a few examples.

$$\frac{2\ mol\ NaI}{1\ mol\ PbI_2} \qquad \frac{2\ mol\ NaI}{2\ mol\ NaNO_3} \qquad \frac{1\ mol\ PbI_2}{1\ mol\ Pb(NO_3)_2}$$

In the laboratory, we measure the number of moles of a substance by measuring the mass of the substance. We use the molecular weight (formula weight for ionic compounds) of the substance as our conversion factor between grams and moles. *A mole is the mass of a substance equal to its molecular weight expressed in grams.* For example, the formula weight of NaI is 150. Therefore, 150 g of NaI is one mole of NaI. Likewise, if we have 75 g of NaI, then we have 0.50 moles of NaI.

Using the stoichiometric factors for a balanced chemical equation and the relationship between mass, molecular weight, and moles, we can perform a variety of stoichiometric calculations with any chemical equation. Here are a couple of examples.

Suppose we want to know how many grams of PbI_2 could be produced by the complete reaction of 13 g of NaI. Here is how we would set up this calculation.

$$\text{\# g } PbI_2 = 13 \text{ g NaI} \times \frac{1 \text{ mol NaI}}{150 \text{ g NaI}} \times \frac{1 \text{ mol } PbI_2}{2 \text{ mol NaI}} \times \frac{461 \text{ g } PbI_2}{1 \text{ mol } PbI_2} = 20 \text{ g } PbI_2$$

| what we are trying to find | what we are given | formula weight for NaI | stoichiometric factor | formula weight for PbI_2 |

The calculation shows us that if we react 13 g of NaI with enough $Pb(NO_3)_2$ for complete reaction, we will produce 20 g of PbI_2. Here are the steps that were followed in performing this calculation.

Step 1: The starting mass of NaI (13 g) is converted to moles of NaI using the formula weight of NaI (150 g/mol). The mass is divided by the formula weight so that the units cancel properly.

Step 2: Moles of NaI are converted to moles of PbI_2 using the stoichiometric factor obtained from the balanced chemical equation.

Step 3: Moles of PbI_2 are converted to grams of PbI_2 using the formula weight of PbI_2 (461 g/mol).

These steps are illustrated in the following general diagram where grams of substance A are converted to grams of substance B.

$$\text{grams A} \xrightarrow{\text{molecular weight of A}} \text{moles A} \xrightarrow{\text{stoichiometric factor}} \text{moles B} \xrightarrow{\text{molecular weight of B}} \text{grams B}$$

Now, suppose we want to calculate the number of grams of $Pb(NO_3)_2$ it would take to produce 15 g of PbI_2. The calculation is similar to the one we performed above and follows the same steps.

$$\text{\# g } Pb(NO_3)_2 = 15 \text{ g } PbI_2 \times \frac{1 \text{ mol } PbI_2}{461 \text{ g } PbI_2} \times \frac{1 \text{ mol } Pb(NO_3)_2}{1 \text{ mol } PbI_2} \times \frac{331 \text{ g } Pb(NO_3)_2}{1 \text{ mol } Pb(NO_3)_2} = 11 \text{ g } Pb(NO_3)_2$$

The calculation shows us that it would take 11 g of $Pb(NO_3)_2$ to produce 15 g of PbI_2 in this chemical reaction.

In today's experiment, you will study the stoichiometric relationship between an acid and a base. The basic salt sodium carbonate (Na_2CO_3) reacts with sulfuric acid (H_2SO_4) to form sodium sulfate (Na_2SO_4) and carbonic acid (H_2CO_3). The carbonic acid immediately decomposes into water and carbon dioxide gas.

$$Na_2CO_{3(s)} + H_2SO_{4(l)} \longrightarrow Na_2SO_{4(s)} + H_2CO_{3(l)}$$

$$H_2CO_{3(l)} \longrightarrow CO_{2(g)} + H_2O_{(l)}$$

When the reactants are mixed together in an open container, the mass of the reacting mixture decreases as carbon dioxide gas escapes into the atmosphere. *The mass of carbon dioxide formed during the reaction can be calculated by determining the amount of mass lost by the reaction mixture.*

The goal of this lab is to study the stoichiometric relationship between sulfuric acid and sodium carbonate. This will be done by incrementally changing the amount of sodium carbonate that is reacted with a fixed amount of sulfuric acid. This creates a situation where, in the early trials, sodium carbonate is the **limiting reactant** and, in the later trials, sulfuric acid is the limiting reactant. When the moles of CO_2 evolved are plotted against the moles of Na_2CO_3 used, it is possible to determine the **stoichiometric equivalence point** for the reaction.

Procedure

The systematic series of trials to be performed are shown in the following table. Note that the mass of sulfuric acid is held constant while the mass of sodium carbonate is being increased with each trial.

Trial	$Na_2CO_3 \cdot H_2O$ Mass in Grams	$3 M H_2SO_4$ Mass in Grams
1	0.24	2.68
2	0.50	2.68
3	1.00	2.68
4	1.48	2.68
5	2.00	2.68
6	2.48	2.68
7	3.00	2.68

Clean and <u>dry</u> a 125 mL Erlenmeyer flask before performing each trial. Record the weight of the flask on the data sheet. Weigh the appropriate amount of sodium carbonate into the flask for the trial being performed. Be certain to keep track of the trial number by writing the number in pencil on the beaker. Record the mass of the sodium carbonate and flask on the data sheet.

Weigh approximately 2.68 g of $3 M H_2SO_4$ in a <u>dry</u> small beaker by taring the beaker and carefully adding drops of sulfuric acid with an eyedropper. Record this mass on the data sheet. Be very careful not to spill any sulfuric acid on the balance. If a spill should occur, clean it up immediately with wet paper towels.

Return to your workbench with the weighed reactants and add the sulfuric acid to the sodium carbonate in the Erlenmeyer flask. Make sure the transfer of acid is complete. Swirl the flask to ensure that the reaction is complete. Allow several minutes for all of the CO_2 gas to evolve. Weigh the flask after all of the carbon dioxide has evolved (the bubbling stops) and enter the result on your data sheet.

From the above measurements, calculate the total mass of reactants before each reaction was initiated by adding the masses of sodium carbonate and sulfuric acid used for each trial. Secondly, determine the total mass of products after the reaction by weighing the 125 mL flask after the reaction has gone to completion. Subtract the weight of the flask to determine the mass of the products. Finally, determine the mass of CO_2 lost in each trial by calculating the difference in the mass of the reaction mixture before and after the reaction. After all seven runs have been performed, complete the data analysis below.

Data Analysis

The information in the following table will help you perform some of the calculations in this experiment.

Formula Weight of $Na_2CO_3 \cdot H_2O$	124.0 g/mole
Moles of $3M H_2SO_4$ per gram	0.003 moles/g

Using the graph on the data sheet, plot **moles of CO₂ produced (evolved or lost)** for each trial on the y-axis and **moles of Na₂CO₃ used** on the x-axis. The rows of values to use to make your graph are highlighted in bold as the last two rows on your data sheet. Check the range of the numbers in these rows in order to determine the appropriate scale to use for the x- and y-axes of your graph. <u>Write the complete balanced equation for the reaction at the top of your graph.</u>

<u>Circle</u> the point (coordinate) where there are stoichiometrically equivalent amounts of both reactants. This is called the **stoichiometric equivalence point** and occurs when both reactants are completely consumed. If the stoichiometry of a reaction is 1:1, at the stoichiometric equivalent point there will be equal moles of both reactants. If the stoichiometry of a reaction is 2:1, at the stoichiometric equivalence point there are two times as many moles of one reactant relative to the other.

The mole ratio of reactants at the stoichiometric equivalence point shown on your graph represents the stoichiometry of the reaction.

Safety Precautions

This lab experiment involves the use of sulfuric acid (H_2SO_4), which may burn your skin and produce holes in your clothes. Wash your hands and work area thoroughly with water and wet paper towels if there is a spill.

Waste Management

All solutions can be poured down the sink and flushed with tap water.

Stoichiometry: Loss of CO_2
Prelab Questions

Name _____ **Section** _____

1. Consider the reaction:

$$3A + 5B + 1C \longrightarrow 2D + 1E$$

If you have 6 moles of reactant A and excess of B and C, how much product E would be formed? (Show your work).

2. For the following metal and non-metal combination reaction you have 3 moles of Na. How many moles of Cl_2 would you need to add to have stoichiometrically equivalent amounts of Na and Cl_2?

$$2 Na_{(s)} + Cl_{2\ (g)} \longrightarrow 2 NaCl_{(s)}$$

Stoichiometry: Loss of CO$_2$ Data Sheet

Name _____ **Section** _____

Name of Lab Partner(s) _____

	Trial 1	Trial 2	Trial 3	Trial 4	Trial 5	Trial 6	Trial 7
Mass of Flask + Salt							
Mass of Flask							
Mass of Salt							
Mass of Acid							
Total Mass of Reactants							
Mass of Products + Flask							
Total Mass of Products							
Mass CO$_2$ Lost							
Moles of Acid Used							
Moles of Salt Used (x-axis)							
Moles of CO$_2$ Lost (y-axis)							

Stoichiometry: Loss of CO₂ Data Sheet

Name _____ Section _____

Graph of CO₂ Lost vs. Na₂CO₃ Used

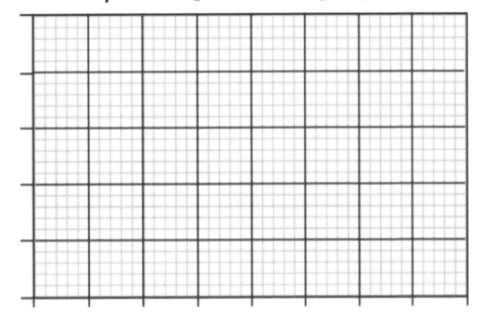

Moles of Na₂CO₃ Used

1. Circle the stoichiometric equivalence point on your graph.

2. What is the stoichiometric ratio for the two reactants? _____

3. In the space below, write the complete balanced equation for the reaction of sodium carbonate with sulfuric acid. Do the coefficients in the balanced equation match the stoichiometric ratio determined in this experiment?

4. a) What was the limiting reactant in trial 2? _____

    b) What was the limiting reactant in trial 6? _____

5. Describe the various sections of your graph and how they relate to the stochiometry of the reaction. In your explanation, discuss your graph as if you were talking to someone unfamiliar with this experiment.

Stoichiometry: Loss of CO_2
Review Questions

Name _____ **Section** _____

1. If 5.00 g of $Na_2CO_3 \cdot H_2O$ (124 g/mole) are combined with 0.010 L of 3 M H_2SO_4, what is the maximum number of moles of CO_2 produced in the reaction? (3 M H_2SO_4 contains 3 moles of H_2SO_4 per liter of solution.)

2. a) Write a balanced equation for the reaction of sodium bicarbonate ($NaHCO_3$) and sulfuric acid (H_2SO_4), yielding sodium sulfate, carbon dioxide, and water.

   b) Make a table and <u>sketch a graph</u> of moles of CO_2 evolved versus moles of $NaHCO_3$ if the sulfuric acid reactant was kept constant at 5 moles and the salt concentration was increased from: 5, 10, 15, and 20 moles.

3. Study the following data and determine the stoichiometry of the reaction of A with B to produce C. Write the balanced equation.

Run	Moles of A	Moles of B	Moles of C
1	0.001	0.01	0.0005
2	0.005	0.01	0.0025
3	0.01	0.01	0.005
4	0.02	0.01	0.01
5	0.04	0.01	0.01
6	0.06	0.01	0.01

Equation:

Stoichiometry: Loss of CO$_2$
Supplemental Questions

Name _____ **Section** _____

Temperature and Reaction Rate

Introduction

In this laboratory exercise, you will use information gathered from the *Kinetics* experiment to study how temperature changes affect the rates of chemical reactions. The relationship between reaction rate and temperature is given by the **Arrhenius equation.**

$$k = Ae^{-Ea/RT}$$

Arrhenius included four variables in this equation; the **rate constant** (k), the **activation energy** (E_a) of the reaction, the temperature in degrees Kelvin (T), and a term known as the **frequency factor** (A). The frequency factor is related to the number of collisions of reactant molecules and the orientation of the molecules. The value of the **ideal gas constant, R,** used in the Arrhenius equation is 8.31 J/mol-K. An examination of the Arrhenius equation shows that increasing the temperature of a reaction causes a corresponding increase in the rate constant, k, which corresponds to a faster rate of reaction.[1]

Taking the natural logarithm of both sides of the Arrhenius equation transforms it into a linear equation where 1/T is the independent variable, *ln k* is the dependent variable, and *ln A* is the intercept.

$$ln\ k = \frac{-E_a}{RT} + ln\ A$$

Therefore, *a plot of ln k vs. 1/T produces a straight line whose slope is E_a/R.* This is the most commonly used method for determining the activation energy (E_a) of a reaction. A sample plot is shown in Figure 1.

In this experiment, you will determine the activation energy of the reaction between BrO_3^- and I^- in acidic solution. This is the same chemical system studied in the *Kinetics* experiment. The balanced equation for this reaction is shown in equation (1).

$$6\ I^- + BrO_3^- + 6\ H^+ \longrightarrow 3\ I_2 + Br^- + 3\ H_2O \qquad \text{[Eqn. 1]}$$

The rate law for this reaction is shown in equation (2).

$$Rate = k\ [BrO_3^-]\ [I^-]\ [H^+]^2 \qquad \text{[Eqn. 2]}$$

---

[1] A catalyst also increases the rate of a reaction, but it does so by lowering the activation energy, E_a. As Ea gets smaller, the reaction rate constant, k, gets larger.

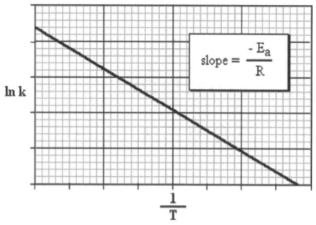

$$\text{slope} = \frac{-E_a}{R}$$

FIGURE 1 — *Arrhenius Equation Plot*

The rate of this reaction is determined experimentally by introducing into the reaction flask a second reaction called a **clock reaction**. A clock reaction is one that changes color when a predetermined amount of reagent has been consumed. The clock reaction used in this experiment is shown in equation (3).

$$I_2 + 2\,S_2O_3^{2-} \longrightarrow 2\,I^- + S_4O_6^{2-} \qquad \text{[Eqn. 3]}$$

The reaction between iodine (I_2) and thiosulfate ($S_2O_3^{2-}$) is nearly instantaneous. As I_2 is produced in reaction (1) it is immediately consumed in reaction (3). When all of the $S_2O_3^{2-}$ is gone, the concentration of I_2 starts to increase in the solution. The addition of starch to the solution produces a deep blue color with the I_2. Therefore, the appearance of a blue color in the reaction flask indicates when a known amount of BrO_3^- has reacted.

Procedure

A total of four trials will be conducted. Each trial will require adding the contents of Flask B to the contents of Flask A and measuring the time it takes for the blue color to appear. Each trial is run in exactly the same manner except for a difference in temperature. Timing of reaction occurs immediately after contents of Flask B are combined into Flask A. Once combined, the reaction in Flask A should be maintained at the designated temperature.

Using your 10 mL graduated cylinder, measure the quantities described in Trial 1 of Table 1 into your designated Flask A and Flask B. Be sure to rinse the graduated cylinder with deionized water before addition of each reagent. Dry the cylinder to avoid diluting the concentrations of the reagents. **Place 4 drops of starch into Flask B.**

Perform the first trial at approximately 0°C. To achieve this low temperature you will need to use 2 large 600 mL beakers as water baths filled with ice and tap water. Place each flask, A and B, in a respective water bath and monitor the temperature of each until the solutions in both flasks are equivalent temperatures. When the solutions are at the desired equivalent temperature, note the time and add the contents of Flask B to Flask A. Swirl the flask continuously and record the time when the solution turns blue. Record the temperature of Flask A on your data sheet. Be sure to rinse the flasks with deionized water before reusing them for the subsequent trials.

Table 1 *Reagent Volumes (mL)*

Flask A (250mL)			Flask B (125mL)	
Molarity: 0.010M KI 0.0010M Na$_2$S$_2$O$_3$ H$_2$O			0.040M KBrO$_3$ 0.10M HCl	
Trial 1	5	5	5	5	5

Repeat the procedure at approximately 10°C, using water baths containing more tap water and less ice than the first trial. Repeat the procedure again at room temperature (approximately 20–25°C), in which water baths are not needed. Finally, perform the reaction at 40°C. To achieve this higher temperature the water baths must be heated slightly. Place the water baths on wire gauze held onto a ring stand by a ring clamp. (Note that the temperature does not have to be these exact values, but you should know exactly what each temperature value is.)

The samples run at lower temperatures may take considerably longer than three minutes to turn blue, but none of them should require more than ten minutes.

Calculations

Determination of Rates

Convert temperature values from Celsius into Kelvin and record these on your data sheet for each trial. Using the determined time values (in seconds) for each trial, record the rate value on your data sheet by dividing the concentration value of 3.3×10^{-5} M by time.

Determination of Rate Constant k

To calculate the rate constant, k, apply the rate and concentration values for each trial in the following equation (remember that concentrations in molarity were determined in the first problem of the Prelab question).

$$\text{Rate} = k\,[BrO_3^-][I^-][H^+]^2 \qquad \text{[Eqn. 2]}$$

Determination of E_a

Convert rate constant, k, values into natural log of k, *ln k*. Convert temperature values (Kelvin) into the reciprocal value, $1/T$ (K^{-1}) for each trial. Record these values on your data sheet.

Using data obtained at all four temperatures, plot *ln k* vs. $1/T$ on your data sheet. Draw a best-fit straight line through the four data points. The slope of this line will be equal to $-E_a/R$, where R = 8.314 J/mol-K. Determine the value of E_a (in Joules/mol) from the slope of the line.

Safety Precautions

Use caution when working with HCl as it is may cause burns.

Waste Management

All waste from this experiment may be flushed down the sink with plenty of water.

Temperature and Reaction Rate
Prelab Questions

Name _____ **Section** _____

1. Using the initial concentration and volume values from Trial 1 of **Table 1**, fill in the following chart of new concentration values (in Molarity) for the following reactants:

Trial	$[BrO_3^-]$	$[I^-]$	$[H^+]$
1			

2. Circle all of the following that change with the temperature of a reaction.

   a) ideal gas constant, R

   b) activation energy, E_a

   c) concentration of reactants, molarity

   d) rate constant, k

   e) order of reaction

3. Draw an arrow on this graph of the Arrhenius equation that shows where you would find *ln A* (the log of the frequency factor)?

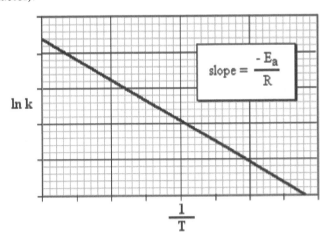

$$\text{slope} = \frac{-E_a}{R}$$

4. If the slope of the line in the graph above is −350 K, what is the value of E_a (in J/mol)?

Temperature and Reaction Rate
Data Sheet

Name _____ Section _____

Name of Lab Partner(s) _____

T (°C)	T (K)	Time (s)	Rate $[3.3\times10^{-5}]/t$	k	1/T	*ln k*

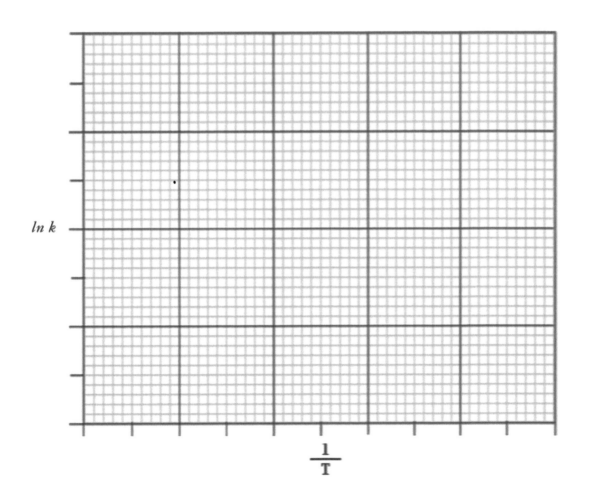

Calculations

Slope = _____ E_a = _____ J/mol

Temperature and Reaction Rate
Review Questions

Name _____ Section _____

1. If 2×10^{-4} mol/L of $S_2O_8^{2-}$ is consumed in 188 seconds, what is the rate of consumption of $S_2O_8^{2-}$?

2. Adding a suitable catalyst to a chemical reaction will (circle all correct choices)

   a) increase the activation energy.

   b) decrease the activation energy.

   c) increase the reaction rate constant.

   d) decrease the reaction rate constant.

   e) increase the rate of the reaction.

   f) decrease the rate of the reaction.

3. The following table lists information taken from a single trial of a kinetics experiment where the reaction is *first order* with respect to reactant A and *zero order* with respect to reactant B. Calculate the reaction rate constant for this reaction.

Trial	[A]	[B]	Initial Rate
1	0.0100 M	0.0200 M	0.150 mol/L-s

Temperature and Reaction Rate
Supplemental Questions

Name _____ **Section** _____

Titrations: Determination of the Molarities of Strong & Weak Acids

Introduction

Titrations are among the most common procedures used in clinical and industrial laboratories. A titration is performed by slowly adding measured amounts of one solution, the **titrant**, usually dispensed from a buret, to another solution until the **equivalence point** is reached. The equivalence point in a titration is where you have added a stoichiometrically equivalent number of moles of titrant to the reactant in the flask. A typical titration apparatus is shown in Figure 1.

The purpose of any titration is to determine the unknown concentration of a solution. This is done by titrating a measured amount of the unknown solution with a known volume of another solution whose concentration we do know. The known solution must be one that chemically reacts with the unknown. If we know the stoichiometry of the reaction between the two reactants, then we can calculate the concentration of the unknown. The only thing missing from our explanation is a method for determining when we have reached the equivalence point of the titration. The specific method for detecting the equivalence point of a titration depends on the nature of the reaction we are studying. For example, a spectrophotometer could be used to detect the equivalence point of a titration in which there was a change in the amount of light absorbed by the solution at the equivalence point. A conductivity meter could be used to detect the equivalence point in a titration that underwent a change in the number of ions in solution at the equivalence point.

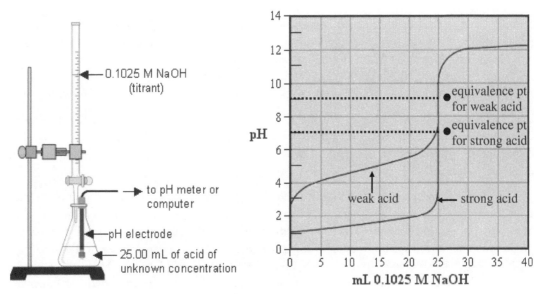

FIGURE 1 — *Titration Apparatus and Titration Curve*

The original technique for estimating the equivalence point of acid-base titrations was the use of **acid-base indicators**. Acid-base indicators are substances that change color when the solution reaches a certain pH. If the properties of the indicator are carefully matched to the conditions that exist at the equivalence point of a titration, the indicator can be a good estimate of the equivalence point. The point in a titration where the indicator changes color is known as the **endpoint** of the titration.

In today's experiment, you will be titrating an acid with a base and following the reaction with the aid of a pH meter (part of the MicroLab data acquisition system). This is referred to as a **potentiometric titration.** Acid-base reactions undergo a sudden change in pH at the equivalence point, so a pH meter makes an excellent tool for determining when the equivalence point of the titration has been reached. By plotting the pH of the reaction mixture vs. the number of milliliters of base added from the buret, a titration curve is obtained that can be used to determine the equivalence point. The equivalence point is in the center of the steeply rising section of the graph. The number of milliliters of base needed to reach the equivalence point is called the **equivalence volume.** Figure 1 shows the titration curves for both a strong acid and a weak monoprotic acid titrated with NaOH. Note the locations of the equivalence points for both titrations and the equivalence volumes of NaOH. All of the acids used in today's experiment are monoprotic acids (they have only one ionizable hydrogen ion). This means that the stoichiometry for the reaction between the acid and NaOH is 1:1.

The following relationship can be used to determine the concentration of a monoprotic acid in an acid-base titration. We will use the data from Figure 1 as an example of this type of calculation.

At the equivalence point of a titration where the stochiometry is 1:1:

$$\text{moles of acid in the flask } = \text{ moles of base added from the buret}$$

$$\text{molarity}_{acid} \times \text{volume}_{acid} = \text{molarity}_{base} \times \text{volume}_{base}$$

$$\text{molarity}_{acid} = \frac{\text{molarity}_{base} \times \text{volume}_{base}}{\text{volume}_{acid}}$$

$$\text{molarity}_{acid} = \frac{0.1025 \text{ M} \times 25.0 \text{ mL}}{25.0 \text{ mL}} = 0.0103 \text{ M}$$

Procedure

Part A: Potentiometric Titration of an Unknown Strong Acid

This experiment can be performed with a pH meter and pH electrode or with a computer and pH electrode. The only significant difference in the procedure is that, if a computer is used, it will automatically plot your titration curve for you. Your instructor will explain any other variations in the procedure that may be required if a computer is used to collect the titration data. Before proceeding, please read **Appendix E** to become familiar with the proper care and use of a pH electrode.

Using a 25 mL volumetric pipet, deliver 25 mL of the unknown strong acid sample into a 100 mL beaker. Do not add any deionized water to this sample. Add 2 drops of phenolphthalein acid/base indicator in order for you to visually determine the endpoint of the reaction. Phenolphthalein is colorless in acidic solution and changes to a pink color in basic solution.

Obtain a 50 mL buret, buret clamp, and small funnel. Assemble the titration apparatus and clean the buret by pouring DI water from a beaker through a funnel at the top of the buret, and then pour a portion of standardized NaOH through the buret. Fill the buret with the standardized NaOH solution using a small funnel to help pour the NaOH. Remove the funnel from the buret. Your instructor will furnish you with the value of the molarity of the NaOH. Put the 100 mL beaker containing the acid under the buret and place the pH electrode into the beaker (refer to Figure 2). Leave the electrode in the beaker for the duration of the titration. Swirl the beaker to get an initial pH value before adding any NaOH titrant. Record this pH value.

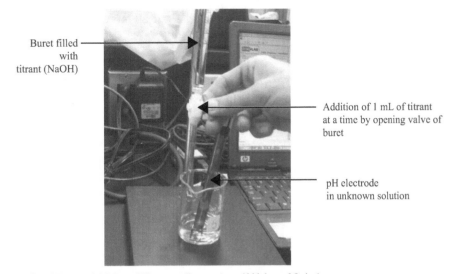

Buret filled
with
titrant (NaOH)

Addition of 1 mL of titrant
at a time by opening valve of
buret

pH electrode
in unknown solution

FIGURE 2—*Addition of Titrant to Determine pH Value of Solution*

Add the first milliliter increment of NaOH from the buret. The electrode can be used to gently stir the solution to maintain equilibrium during the titration. Record the pH value on the table provided in the data sheet. Continue adding NaOH solution in 1 mL increments and enter the corresponding volume of NaOH used after each increment. When a pink color begins to appear and then disappear in the solution in the beaker, begin adding NaOH in 0.5 mL increments and recording the corresponding pH values. When the equivalence point has been reached and the solution remains pink, immediately remove the electrode from the strong basic solution, rinse it well with DI water and store it. *Strongly basic solutions will destroy the glass of the electrode.*

Create a graph of pH vs. Volume of NaOH. Find the equivalence point of the titration from the graph. The equivalence point will be located approximately in the middle of the steep section of your graph. Record the following information on your data sheet: pH when the phenolphthalein turned pink, pH at the equivalence point, and equivalence point volume of NaOH. With this information, you can now calculate the molarity of the unknown strong acid. On your graph, label the equivalence point, equivalence volume, and pH at the equivalence point.

Part B: Potentiometric Titration of Vinegar

In this part of the experiment, you will titrate a weak acid (acetic acid) and compare the results with those you obtained for the titration of a strong acid. Commercial vinegar is approximately 5% acetic acid and will be used as the source of acetic acid in this experiment. The structure of acetic acid is shown below:

$$H-\overset{\overset{\textstyle H}{|}}{\underset{\underset{\textstyle H}{|}}{C}}-\overset{\overset{\textstyle O}{\|}}{C}-O-H$$

The –COOH group is known as the **carboxylic acid** functional group. Acetic acid is a weak acid that exists in equilibrium with its conjugate base in aqueous solution, as shown in Equation 1. The **equilibrium expression** for K_a for acetic acid is shown in Equation 2.

$$CH_3COOH + H_2O \rightleftharpoons H_3O^+ + CH_3COO^- \qquad \text{[Eqn. 1]}$$

$$K_a = \frac{[H_3O^+][CH_3COO^-]}{[CH_3COOH]} \qquad \text{[Eqn. 2]}$$

Acetic acid reacts completely with a strong base to form water and sodium acetate as shown in the following equation.

$$NaOH_{(aq)} + CH_3COOH_{(aq)} \longrightarrow H_2O_{(l)} + Na^+_{(aq)} + CH_3COO^-_{(aq)}$$

The presence of the acetate ion, a weak base, at the equivalence point of the acetic acid titration explains why the pH at the equivalence point is greater than 7.

During the course of this titration, there is a point reached where exactly half of the acetic acid has been neutralized with NaOH. At this point, the concentration of acetate ion is exactly equal to the concentration of unreacted acetic acid. Equation 2 above shows that under these circumstances, the K_a of acetic acid is equal to the H_3O^+ concentration in the solution. This point in the titration occurs at the half-equivalence volume. Therefore, at the half-equivalence volume of the titration, the following relationship exists.

$$[H_3O^+] = K_a \text{ (only true at the half-equivalence volume of the titration)}$$

Taking the -log of both sides of the equations leads to the following relationship.

$$pH = pK_a \text{ (only true at the half-equivalence volume of the titration)}$$

To determine the K_a of acetic acid, find the pH at the half-equivalence volume and use the following formula to convert pK_a to K_a.

$$K_a = 10^{-pK_a}$$

Using a 5.00 mL volumetric pipet, add 5.00 mL of vinegar to a 100 mL beaker and then add 50 mL of DI water to the vinegar in the beaker. Add 2 drops of phenolphthalein indicator so that you will be able to visually determine the endpoint of the titration.

Place the 100 mL beaker containing the acid under the buret and place the pH electrode into the beaker. Leave the electrode in the beaker for the duration of the titration. Swirl the beaker to get an initial pH value before adding any NaOH titrant. Record this pH value.

Add the first milliliter increment of NaOH. The electrode can be used to gently stir the solution to maintain equilibrium during the titration. Record the pH value on the table provided in the data sheet. Continue adding NaOH solution in 1 mL increments and enter the corresponding volume of NaOH used after each increment. When a pink color begins to appear and then disappear in the solution in the beaker, begin adding NaOH in 0.5 mL increments and recording the corresponding pH values. When the equivalence point has been reached and the solution remains pink, immediately remove the electrode from the strong basic solution, rinse it well with DI water and store it. *Strongly basic solutions will destroy the glass of the electrode.* When your experiment is complete, rinse the buret with several portions of DI water poured from a beaker, ensuring that the DI water runs through the tip of the buret into a spare beaker.

Create a graph of pH vs. Volume of NaOH. Find the equivalence point of the titration from the graph. Record the following information on your data sheet: pH when the phenolphthalein turned pink, pH at the equivalence point, equivalence point volume of NaOH, and the pH at the half-equivalence volume. With this information, you can now calculate the molarity of the acetic acid and the value of K_a. On your graph, label the equivalence point, equivalence volume, the pH at the equivalence point, the half-equivalence volume, and the pH at the half-equivalence volume.

Safety Precautions

Use caution when working with NaOH solutions. They are corrosive and can cause serious eye injuries.

Waste Management

All solutions may be flushed down the sink with plenty of water. Do not pour any unused NaOH solution from your buret into the original reagent bottle, as this may cause contamination.

Titrations: Determination of the Molarities of Strong & Weak Acids Prelab Questions

Name _____ **Section** _____

1. Give a balanced equation for the neutralization reaction that occurs in the reaction of a strong base, NaOH, with a strong acid, HCl.

2. In this experiment, what solution is placed in the buret to perform titration of an acid?

3. When is the pH equal to the pK_a for a weak acid in an acid base titration?

Titrations: Determination of the Molarities of Strong & Weak Acids Data Sheet

Name _____ Section _____

Name of Lab Partner(s) _____

Part A: Titration of 25.00 mL of Unknown Strong Acid

Volume of NaOH (mL)	pH

Part A: Titration of an Unknown Strong Acid

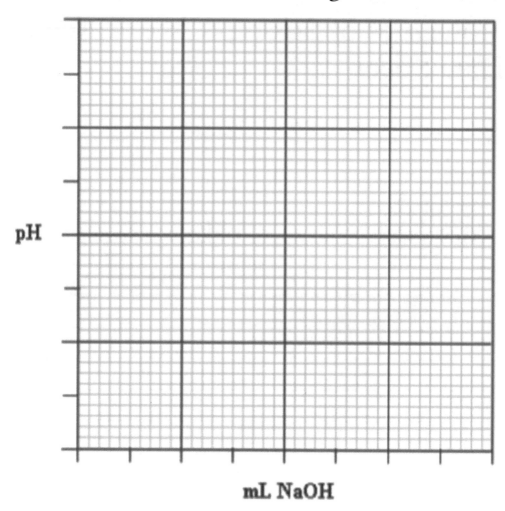

pH

mL NaOH

What was the approximate pH when the solution remained pink? _____

What is the pH at the equivalence point, as determined by the graph? _____

What is the equivalence volume, as determined by the graph? _____

Calculations of molarity of unknown strong acid:

Molarity of unknown strong acid _____

Part B: Titration of 5.00 mL of Acetic Acid, a Weak Acid

Volume of NaOH (mL)	pH

Part B: Titration of a Weak Acid, Acetic Acid

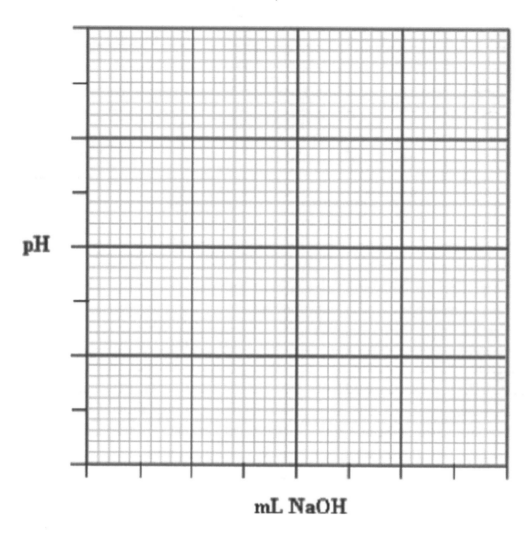

mL NaOH

What was the approximate pH when the solution remained pink? _____

What is the pH at the equivalence point, as determined by the graph? _____

What is the equivalence volume, as determined by the graph? _____

Calculation of molarity of acetic acid:

Molarity of acetic acid _____

What is the pK_a of acetic acid? (Remember, pK_a is equal to the pH at half the equivalence volume) _____

What is the K_a of acetic acid? _____

Titrations: Determination of the Molarities of Strong & Weak Acids Review Questions

Name _____ **Section** _____

1. List at least three differences you observed between the titration graphs of the strong and weak acids. Explain these differences.

   a)

   b)

   c)

   d)

2. What volume of 0.105 M NaOH is required to titrate each of the following solutions to the equivalence point? In each case indicate whether the equivalence point is acidic, basic or neutral.

   a) 20 mL of 0.075 M HNO_3

   b) 15 mL of 0.165 M formic acid (HCO_2H)

3. Two monoprotic acids, HX and HY, both at 0.1 M concentration, are titrated with 0.10 M NaOH. The pH at the equivalence point for HX is 8.8, while that for HY is 7.9. Which is the weaker acid?

Titrations: Determination of the Molarities of Strong & Weak Acids Supplemental Questions

Name _____ **Section** _____

Unknown Solid Identification

Introduction

In the experiment titled *Exchange Reactions*, you had to identify five solutions on the basis of their physical and chemical properties. Physical properties such as color, odor, and acidity or basicity, and chemical properties such as chemical reactivity were used to determine the identities of your five unknowns. In today's experiment you are to identify a single solid substance from a large list of possibilities. You will be using many of the same techniques you learned in *Exchange Reactions*; therefore you will need to refer to the information in the *Exchange Reactions* experiment often. Several new techniques will be introduced in this experiment to help you identify your unknown.

The unknown solid that you receive is one from the list of 25 compounds shown below. To aid you in identifying your unknown, these compounds are classified into five categories, with some compounds belonging to more than one category. The categories and compounds are shown below.

Acidic	Basic	Halides	Nitrates	Carbonates
NH_4Cl	Na_2CO_3	NH_4Cl	$NaNO_3$	Na_2CO_3
$ZnCl_2$	K_2CO_3	KI	KNO_3	K_2CO_3
$Zn(NO_3)_2$	$(NH_4)_2CO_3$	KCl	$Ca(NO_3)_2$	$(NH_4)_2CO_3$
$Cu(NO_3)_2$	Na_2HPO_4	KBr	$Zn(NO_3)_2$	
$CuSO_4$	K_2CrO_4	NaI	$Cu(NO_3)_2$	
$Ni(NO_3)_2$	$(NH_4)_2CrO_4$	$NaBr$	$Fe(NO_3)_3$	
$Fe(NO_3)_3$		$NaCl$	$Ni(NO_3)_2$	
$CuCl_2$		$CaCl_2$	$LiNO_3$	
NH_4Br		$CaBr_2$		
		$ZnCl_2$		
		$CuCl_2$		

All of the compounds in the **acidic** and **basic** categories will change the color of litmus paper. (Compounds should <u>not</u> be identified based on acidity/basicity alone, as this is the least reliable of the confirmatory tests). All of the compounds in the **halide** category will react with $AgNO_3$ and $Pb(NO_3)_2$ to produce precipitates. Substances in the **nitrate** category must be identified on the basis of the cation alone, because the nitrate ion is very difficult to identify. The ions of sodium, potassium, and calcium produce distinctive colors when heated in a flame. Compounds in the **carbonate** category will react with acids to produce bubbles of CO_2.

You will need to perform a number of different tests in order to conclusively identify your compound. After you have determined which compound you have, you will write a one-page report describing the reasons for your identification.

Procedure

Fill a medium test tube to an approximate depth of 2 cm with one of the 15 solid unknowns assigned to you by your instructor. Add 3 mL of deionized water to a second medium test tube. Place a small amount of your unknown solid in the test tube containing the water. Stopper the test tube and shake it until the entire solid dissolves. Repeat this process until no more solid dissolves in the solution. The solution is now saturated. This is the solution you are to use whenever the procedure calls for you to use the solution of your unknown. The tests you are to perform on your unknown are outlined below.

Test 1: Observe the color of the ***unknown solid*** and ***solution***. Record your observations on the data sheet.

Test 2: Observe the odor of the ***unknown solid*** and ***solution*** by wafting vapors from the test tube toward your nose. Record your observations on the data sheet.

Test 3: Place 10 drops of your ***unknown solution*** in a medium test tube. Add 10 drops of 1 M NaOH to the test tube. Gently warm the solution in a cool burner flame for a few seconds. Do not let the solution boil. Waft vapors from the test tube toward your nose. Record any odors you observe and any changes in the appearance of the solution after it is heated. If the solution contains the NH_4^+ ion, the odor of ammonia should be detected.

$$NaOH + NH_4^+Y^- \longrightarrow NH_4OH \text{ (ammonia odor)} + Na^+Y^-$$

Test 4: Test the ***unknown solution*** with both red and blue litmus to determine its acid/base properties. Recall that blue litmus turns red in acidic solution and red litmus turns blue in basic solution.

Test 5: Many atoms will emit light in the visible spectrum when they are vaporized in a flame. The color of light emitted is characteristic of a particular element. Therefore, observing the color emitted by your unknown when it is heated in a flame can be useful in its identification. Some examples are shown below.

> Na = bright yellow flame
> K = pale violet flame
> Li = bright red flame
> Ca = yellow-orange flame

Obtain a flame test loop and heat it in the hottest part of your flame until the wire glows red. Immediately plunge it into a small test tube containing 1 mL of 6 M HCl to clean it. Repeat this process until heating the test loop does not produce a color in the flame when it is heated to redness. Place a <u>small</u> crystal of your ***unknown solid*** in the loop of the flame test wire. The solid crystal will stay on the loop easily if the flame test loop is wet with HCl. Place the test loop with the crystal in the hottest part of the burner flame (the tip of the inner blue cone of the flame) and record any colors you see produced in the flame. It is very difficult to get sodium ions off the flame test loop and sodium ions are contaminants in most chemicals. Therefore, you may see a little of the sodium flame test when performing the tests for potassium, calcium, and others, including ammonium. Use the fragile test wire carefully and return it promptly upon completion of your tests.

<u>Alternate Flame Test Method</u>: Obtain a wooden splint and soak it in the test tube containing the supersaturated ***solution of your unknown***. When you are ready for the flame test, wave the splint through the flame and observe

the color emitted. Do not hold the sample in the flame as this will cause the splint to ignite. Immediately after the splint is waved through the flame, place it in a beaker of water to extinguish any possible flame. Use a new split for a new test. Discard the splint when lab is complete.

Test 6: Place a small amount of your **unknown solid** in a medium test tube. Add 10 drops of 6 M HCl to the test tube. Record your observations. If the solution bubbles, then CO_2 is being given off. This is evidence of a carbonate-containing compound:

$$X^+_2 CO_3^{-2} + 2\,HCl \longrightarrow CO_{2\,(g)} + H_2O + 2\,X^+Cl^-$$

Test 7: Add 2 drops of 0.1 M $AgNO_3$ to 5 drops of your **unknown solution** in a spot plate well. Record any precipitate formed. Silver ion will form the following precipitates with halides (chloride, bromide, and iodide):

$$AgBr_{(s)} \quad \text{– pale yellow}$$
$$AgCl_{(s)} \quad \text{– white}$$
$$Ag\,I_{(s)} \quad \text{– pale yellow}$$
$$Ag_2CO_{3\,(s)} \quad \text{– yellow-white or gray-white}$$

Test 8: Add 2 drops of 0.1 M $Pb(NO_3)_2$ to 5 drops of your **unknown solution** in a spot plate well. Record any precipitate formed. Lead(II) ions will form precipitates with halides (chloride, bromide, and iodide):

$$PbBr_{2\,(s)} \quad \text{– white}$$
$$PbCl_{2\,(s)} \quad \text{– white}$$
$$PbI_{2\,(s)} \quad \text{– bright yellow}$$
$$PbCO_{3\,(s)} \quad \text{– white}$$

Some of the confirmatory tests you perform may give negative results (no reaction). A negative result is still important information to record on your data sheet. For example, if your unknown gives no precipitate when mixed with Ag^+ or Pb^{2+} ions, then you do <u>not</u> have halide ions (Cl^-, Br^-, or I^-) or carbonate ions (CO_3^{2-}) present in your unknown.

Important Note: The flame tests for sodium and calcium are often difficult to distinguish. The yellow flame of sodium can be easily mistaken for the yellow-orange flame of calcium. To differentiate between Na^+ and Ca^{2+}, add 2 drops of 0.1 M Na_2CO_3 to 2 drops of your unknown solution. A white $CaCO_3$ precipitate will form if your unknown contains Ca^{2+}. No precipitate forms with Na^+.

Once you have identified your unknown, you must write a one-page report using grammatically correct, well-formed sentences explaining the reasons for your identification. This report must include:

A.   Three firm reasons for your identification.

B.   **Balanced chemical equations** for all positive tests used in your identification.

C.   A description of your unknown from the CRC *Handbook of Chemistry and Physics*. You must include the edition of the *Handbook* and the page number of your compound in the report.

Safety Precautions

Be cautious when using the NaOH and HCl solutions. These are fairly concentrated solutions and may cause burns.

Waste Management

Pour wastes from Tests #7 and #8 into the "Metals waste" container in the fume hood. Solid wastes should be placed in the nearest garbage container. Solid wastes should not be placed with the liquid "Metals wastes". All other products from other tests may be flushed down the sink with plenty of water. Do not place unknown solids back into the original containers, as this may introduce contamination.

Unknown Solid Identification
Prelab Questions

1. Sodium ion (Na^+) and calcium ion (Ca^{2+}) produce nearly the same color in a flame test (yellow and yellow-orange, respectively). Describe a way to differentiate between the two using a solution of Na_2CO_3 and write the correct balanced equation(s). (Hint: recall the solubility rules for Na^+ ion.)

2. A saturated solution of an unknown solid reacted with $AgNO_3$ to give a pale yellow precipitate, and also reacted with $Pb(NO_3)_2$ to give a white precipitate. Based on only these two test results, you can conclude that the unknown solid contains which anion? Write the correct balanced equations for reaction of the anion with each of these reagents.

Unknown Solid Identification Data Sheet

Name _____ **Section** _____

Name of Lab Partner(s) _____

Number of Unknown _____

Results

1. Observation of **color** of unknown.

    a) Solid _____

    b) Solution _____

2. Observation of **odor** of unknown.

    a) Solid _____

    b) Solution _____

3. Odor and appearance of solution after addition of **NaOH**.

    a) odor _____

    b) appearance of solution _____

4. **Acid/Base properties** of unknown solution.

5. Results of **flame tests** performed on solid unknown.

6. Reaction of **HCl** with unknown solid.

7. Reaction of **AgNO$_3$** with unknown solution.

8. Reaction of **Pb(NO$_3$)$_2$** with unknown solution.

Identity of Unknown _____

Unknown Solid Identification
Review Questions

1. The colorless solution made from an unknown white solid turned blue litmus red. The flame test showed no color. Addition of NaOH produced a white, gelatinous precipitate but no odor. Addition of HCl produced no bubbles. No precipitate formed after adding $AgNO_3$ and $Pb(NO_3)_2$. Identify the unknown from the list of possibilities in the Introduction to this experiment.

2. The colorless solution of an unknown solid turned red litmus blue. No odor was noted upon adding NaOH. A gray-white precipitate formed upon adding $AgNO_3$, and a white precipitate formed upon adding $Pb(NO_3)_2$. The flame test gave a violet color, and bubbles formed after adding HCl. Identify the unknown from the list of possibilities in the Introduction to this experiment.

3. The colorless solution of an unknown solid turned blue litmus red and emitted an odor of ammonia upon addition of NaOH. A pale yellow precipitate was observed upon reaction with $AgNO_3$ and a white precipitate was observed upon addition of $Pb(NO_3)_2$. Bubbles were not evident upon addition of HCl. Identify the unknown from the list of possibilities in the Introduction to this experiment.

Unknown Solid Identification
Supplemental Questions

Name _____ **Section** _____

Voltaic Cells

Introduction

Chemical reactions that involve an exchange of electrons between reactants are called **oxidation-reduction reactions** or **redox reactions**. The reactant that loses one or more electrons is said to be **oxidized**, while the reactant that gains the electrons is said to be **reduced.** In the spontaneous redox reaction of Ag^+ with copper metal (Cu),

$$2\ Ag^+_{(aq)} + Cu_{(s)} \longrightarrow 2\ Ag_{(s)} + Cu^{+2}_{(aq)}$$

electrons are transferred from the copper metal to Ag^+, resulting in the oxidation of copper metal to copper(II) ions and the reduction of silver(I) ions to silver metal. If a solid piece of copper is introduced into a solution containing silver ions, the exchange of electrons will take place directly as the silver ions collide with the copper metal. However, if these reactants are in separate containers where direct contact is impossible, the spontaneous reaction can still take place if a path for the electron transfer can be provided. This is the principle behind the voltaic cell or battery. In the voltaic cell shown in Figure 1, metal electrodes in the reactant compartments are connected by a wire through which the electrons, released when Cu is oxidized to Cu^{2+}, can travel to the compartment containing Ag^+ ions and reduce them to Ag. In this way, the free energy (ΔG) released by the spontaneous redox reaction can be harnessed as electrical energy.

The electrode at which the oxidation reaction occurs (the Cu electrode in this example) is called the **anode**. In a voltaic cell it is designated by a negative sign (–). The reduction reaction occurs at the **cathode** (the Ag electrode in this example). In a voltaic cell it is designated by a positive sign (+). Electrons flow through the external circuit from the anode to the cathode and ions flow through the salt bridge to keep the two half-cells electrically neutral.

A simple short-hand method has been developed for representing cells that does not depend on drawing elaborate diagrams whenever a voltaic cell must be described. In this notation, the cell shown in Figure 1 can be represented as

$$Cu \mid Cu^{2+} \parallel Ag^+ \mid Ag$$
$$\text{Anode} \quad \text{Cathode}$$

The components of the anode compartment are written on the left and the components of the cathode compartment are written on the right. The components of the two half-cells are separated by a salt bridge, represented by the double vertical line. The single vertical lines represent the phase boundary between the solid electrode and the ions in solution. This notation will be used to describe the reactions you will study in this experiment.

Although no oxidation can take place without a corresponding reduction (and vice versa), it is often convenient to represent the processes occuring at the anode and cathode in a voltaic cell as separate **half reactions**, one

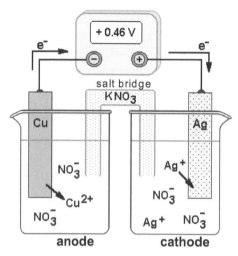

FIGURE 1 — *Voltaic Cell for the Reaction of Copper and Silver*

representing oxidation and the other representing reduction. Since silver ion is being reduced in this example, the reduction half reaction is

$$Ag^+_{(aq)} + e^- \longrightarrow Ag_{(s)}$$

and because copper metal is being oxidized, the oxidation half reaction is

$$Cu_{(s)} \longrightarrow Cu^{2+}_{(aq)} + 2e^-$$

When the reduction half reaction is doubled so that it consumes the same number of electrons as are produced by the oxidation half reaction (two in this case), the two half reactions can be combined to give the overall balanced chemical equation for the reaction.

$$2\,Ag^+_{(aq)} + 2e^- \longrightarrow 2\,Ag_{(s)}$$
$$\underline{Cu_{(s)} \longrightarrow Cu^{2+}_{(aq)} + 2e^-}$$
$$2\,Ag^+_{(aq)} + Cu_{(s)} \longrightarrow 2\,Ag_{(s)} + Cu^{+2}_{(aq)}$$

Because voltaic cells provide free energy in the form of electrical energy, they provide a convenient means to measure the free energy change (ΔG) of a redox reaction. The **electromotive force** (abbreviated **emf**) for the cell, symbolized with a capital **E** with units of **volts,** can be measured directly with a voltmeter. The emf of a reaction is related to the ΔG of the reaction by the following expression

$$\Delta G = -\,nFE$$

where n is the number of electrons exchanged in the reaction and F is a constant known as Faraday's constant (96,500 C/mol). If the emf is measured under standard conditions (1 M concentrations for all solutions, 1 atm pressure for all gases), then it is given the symbol **E°**.

The emf of a complete cell is given by the standard reduction potential of the cathode reaction $E°_{red\,(cathode)}$ <u>minus</u> the standard reduction potential of the anode reaction $E°_{red\,(anode)}$:

$$E°_{cell} = E°_{red\,(cathode)} - E°_{red\,(anode)}$$

For example, the standard reduction potential for Cu^{2+} ions to copper metal is +0.34 V and the standard reduction potential for Ag^+ ions to silver metal is +0.80 V. Therefore the $E°_{cell}$ for the reduction of Ag^+ ions by copper metal is +0.80 V – (+0.34 V) = 0.46 V.

It is common practice to list tables of electrodes potentials as **reduction potentials**. Your textbook has an extensive table of standard reduction potentials.

In this experiment, you will determine the reduction potentials for six half-reactions by connecting them to the $Zn|Zn^{2+}$ half-cell. The standard reduction potential for Zn^{2+} is $-0.76V$.

Procedure

Part A:

The cell potentials for a variety of voltaic cells will be measured using the apparatus shown in Figure 2. This apparatus uses a crucible which has a porous surface on the bottom that substitutes for the salt bridge shown in Figure 1. This lowers the internal resistance of the cell and allows accurate cell potentials to be measured. The zinc half-cell will be the anode (oxidation half-reaction) in all of your trials.

Attach a voltage probe to the MicroLab data acquisition unit. Open the folder labeled Voltaic Cells or the folder designated MicroLab 5.5 software (for use with the newer FS–522 MicroLab units). Open the program labeled Electrochemistry and click on the "Voltaic Cells experiment CHML 102". You will be able to read voltages in real time displayed in the lower right hand corner of the computer screen using this program. Make sure that the red and black leads are terminated with alligator clips. Obtain a zinc electrode and a nickel electrode from your instructor.

Attach the black lead from the voltage probe to the zinc electrode. Attach the red lead from the voltage probe to the nickel electrode. Pour about 10–12 mL of 0.1 M $NiSO_4$ in the porous crucible. Pour a fresh portion of 10–12 mL of 0.1 M $ZnSO_4$ solution into a 100 mL beaker. Place the zinc electrode in the zinc(II) sulfate solution and the nickel electrode in the nickel(II) sulfate solution. Now, carefully lower the porous crucible into the 100 mL beaker. Be sure that none of the $NiSO_4$ in the crucible spills over into the beaker. Wait 15 seconds, then read the cell potential from the MicroLab unit. Record this value on your data sheet.

Clean the beaker, the electrode, and the crucible thoroughly with deionized water. Repeat the experiment by substituting solutions of silver(I) nitrate ($AgNO_3$), lead nitrate ($Pb(NO_3)_2$), copper(II) sulfate ($CuSO_4$), iodine water (I_2) and bromine water (Br_2) in the porous crucible. Use fresh portions of 10–15 mL of zinc(II) sulfate solution in the 50 mL beaker. Refer to the data sheet for the proper combinations of solutions and electrodes for each cell.

After you have measured the six cell potentials described above and recorded these on your data sheet, you are now ready to calculate the standard reduction potentials (E_{red}) for each of the species shown on the data sheet. Complete the table of standard reduction potentials shown in Part A of the data sheet by writing all of the half-cell reactions as reductions.

Part B:

In this part of the experiment you are to predict the value of E_{cell} for each of the following voltaic cell reactions using the data from the table you created in Part A. Be sure to note which species are being used at the anode and cathode. After predicting the cell potentials, measure the actual cell potential using the same technique employed in Part A.

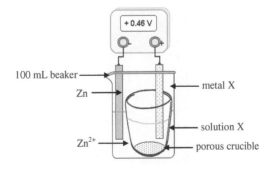

FIGURE 2 — *Voltaic Cell Constructed from Porous Crucible*

$$\text{Pb} \mid \text{Pb}^{2+} \parallel \text{Cu}^{2+} \mid \text{Cu} \qquad\qquad \text{[Eqn. 1]}$$

$$\text{Cu} \mid \text{Cu}^{2+} \parallel \text{Br}_2, \text{Br}^- \mid \text{graphite} \qquad\qquad \text{[Eqn. 2]}$$

Safety Precautions

Caution: Bromine and iodine evolve toxic fumes. Handle these solutions as close to your auxillary exhaust vent as possible. Never leave these reagent bottles open. Wash your hands thoroughly after finishing this laboratory experiment to remove all residual chemicals.

Waste Management

Pour all solutions into the "Metals waste" container in the fume hood. Rinse the metal electrodes with DI water dry them, and return them to the proper containers (petri dishes) on the reagent shelf.

Voltaic Cells Prelab Questions

Name _____ **Section** _____

1. Write a chemical equation for the reaction that occurs in the following cell.

    $$Mg|Mg^{2+}||Ag^+|Ag$$

2. Identify which half-cell reaction is occurring at the anode and cathode respectively in Question #1 above. What signs are used to designate the anode and cathode?

Voltaic Cells Data Sheet

Name _____ **Section** _____

Name of Lab Partner(s) _____

Part A:

Cell	E cell (volts)
a. $Zn \mid Zn^{2+} \parallel Ni^{2+} \mid Ni$	
b. $Zn \mid Zn^{2+} \parallel Ag^{+} \mid Ag$	
c. $Zn /\mid Zn^{2+} \parallel Pb^{2+} \mid Pb$	
d. $Zn \mid Zn^{2+} \parallel Cu^{2+} \mid Cu$	
e. $Zn \mid Zn^{2+} \parallel I_2, I^- \mid graphite$	
f. $Zn \mid Zn^{2+} \parallel Br_2, Br^- \mid graphite$	

Table of Reduction Potentials

Reduction Half-Reactions	E_{red} (volts)
1. $Zn^{2+} + 2e^- \longrightarrow Zn$	-0.76
2.	
3.	
4.	
5.	
6.	
7.	

Question 1. From your results in Part A, give an example of a good oxidizing agent and verify with an explanation _____.

Question 2. Name an example of a good reducing agent_____. Explain.

Part B:

Cell	Predicted E	Measured E
1. Pb | Pb²⁺ || Cu²⁺ | Cu		
2. Cu | Cu²⁺ || Br₂, Br⁻ | graphite		

Question 3. In each cell in Part B, write each half cell reactions and identify which substance is at the anode and the cathode.

Question 4. Include your calculations below:

Voltaic Cells Review Questions

1. Given the following $E°_{red}$ values, calculate the standard cell potential E_{cell} for the spontaneous reaction that would occur between magnesium and silver.

$$Mg^{2+}_{(aq)} + 2e^- \longrightarrow Mg_{(s)} \quad E°_{red} = -2.37 \text{ V}$$

$$Ag^+_{(aq)} + e^- \longrightarrow Ag_{(s)} \quad E°_{red} = +0.80 \text{ V}$$

2. Using the table below, select the best oxidizing agent and the best reducing agent.

Metal	Reduction Potential
A	–0.90 V
B	–0.10 V
C	+0.85 V

Best oxidizing agent: _____   Best reducing agent: _____

3. Draw a diagram of a voltaic cell where $Ti_{(s)}$ is oxidized in one compartment and Cd^{2+} is reduced in the other compartment. Make sure you list the anode and cathode with appropriate signs, show the direction of electron flow outside the cell, and ion flow within the cell. Which is the oxidizing agent and which is the reducing agent in the cell you have drawn?

$$Ti_{(s)} + Cd^{2+}_{(aq)} \longrightarrow Ti^{2+}_{(aq)} + Cd_{(s)}$$

4. Use the configuration of $A|A^+||B^+|B$ to represent the cell you have drawn in Question #3.

Voltaic Cells Supplemental Questions

Name _____ **Section** _____

Water Hardness: Determination with EDTA

Introduction

Water sources are often characterized by how they react with soap. Water which enables soap to dissolve, lather, and sequester dirt is described as **soft water**. **Hard water**, in contrast, contains polyvalent ions such as Ca^{2+}, Mg^{2+}, and Fe^{3+}, which form insoluble precipitates with soap, nullifying its foaming and cleansing action. Hard water is formed when ground water percolates through limestone ($CaCO_3$) or dolomite ($CaCO_3$–$MgCO_3$) rocks or when water dissolves calcium sulfate, magnesium sulfate, or iron salts.

Because hard water forms calcium, magnesium, and iron(III) precipitates with soap, it often causes problems in laundering and in steam generation in boilers. Millions of dollars are spent every year to "soften" water in order to avoid soap scum and boiler scale. Water softening usually consists of replacing Ca^{2+} and Mg^{2+} ions with Na^+ ions. Sodium ions do not react with soap or form boiler scale. However, British researchers found a higher incidence of heart disease in households that use soft water for drinking due to the increased levels of Na^+ in soft water. Their findings indicate that you can save money and lower your risk of heart disease by not attaching your water softener to drinking water pipes.

The objective of today's experiment is to determine the hardness of two different water samples using the classical titration method. All of the polyvalent ions responsible for water hardness are usually lumped together as an equivalent quantity of $CaCO_3$ because the analysis method you will use does not distinguish between the different ions Therefore, even though both Ca^{2+} and Mg^{2+} are responsible for making water hard, you will report your values for water hardness in **parts per million (ppm) of $CaCO_3$**. A concentration of 1 **ppm of $CaCO_3$** in water is the same as **1 mg of $CaCO_3$ per liter** of water. (A liter of water is assumed to be 1000 g at room temperature.)

Water hardness is determined by **titrating** a water sample with a substance called ethylenediaminetetraacetic acid, or simply **EDTA**. EDTA binds to Ca^{2+} and Mg^{2+} in a process called **chelation**. The term chelate is derived from the Greek word meaning "claw", which describes the way EDTA grasps Ca^{2+} and Mg^{2+} in a sort of pincer action. The pH of the solution is adjusted to approximately pH 10 using a buffer solution. At this pH, EDTA is present as a divalent anion written as H_2EDTA^{2-}. This form of EDTA is highly selective toward Ca^{2+} and Mg^{2+} and has little tendency to react with other metal ions that might be in the solution such as Fe^{3+}.

The reaction of H_2EDTA^{2-} with Ca^{2+} is shown in the following equation. The equation is the same for the reaction of H_2EDTA^{2-} with Mg^{2+}.

$$Ca^{2+}_{(aq)} + H_2EDTA^{2-}_{(aq)} \longrightarrow CaEDTA^{2-}_{(aq)} + 2 H^+_{(aq)}$$

The stoichiometry of this reaction is 1:1, which means that 1 mole of EDTA is used for each mole of Ca^{2+} or Mg^{2+} ions present in the sample. In this experiment, you will slowly add a solution of EDTA to a hard water sample (this is called titrating the sample) until you have added as many moles of EDTA as there are moles of Ca^{2+} and Mg^{2+} in the sample. This point in the titration is known as the **equivalence point**. If the concentration of the EDTA solution is known, and if the volume of EDTA solution required to reach the equivalence point is known, then a simple calculation allows you to determine the total concentration of calcium and magnesium ions in the sample.

All of the reactants and products of this reaction are colorless, which means that there is no visual cue telling you when you have reached the equivalence point. Fortunately, there is a special **indicator** available, **Eriochrome Black T**, that changes color near the equivalence point of this titration. The point during a titration where the indicator changes color is called the **endpoint**. If the titration is carefully performed, the endpoint and the equivalence point should be very close to one another. However, it is important to keep in mind that in any titration *the endpoint is always an estimate of the true equivalence point.*

Eriochrome Black T (abbreviated **In** for indicator), which is normally blue, forms a stable complex ion with Mg^{2+} that is red.

$$Mg^{2+} + In \longrightarrow In\text{-}Mg^{2+}\, complex$$
$$\textbf{(blue)} \qquad \textbf{(red)}$$

In this experiment, a small amount of Mg^{2+} and Eriochrome Black T are added to the hard water sample to be titrated. The solution turns red due to the formation of the $In\text{-}Mg^{2+}$ complex. When the hard water sample is titrated with EDTA, the EDTA first reacts with the "free" calcium and magnesium ions in solution. Once all of the "free" calcium and magnesium ions have reacted with EDTA (the equivalence point), the EDTA reacts with the small amount of Mg^{2+} attached to the Eriochrome Black T. When all of this Mg^{2+} has reacted with EDTA, the solution turns blue (the endpoint).

$$H_2EDTA^{2-}_{(aq)} + In\text{-}Mg^{2+}_{(aq)} \longrightarrow MgEDTA^{2-}_{(aq)} + In_{(aq)} + 2\,H^{+}_{(aq)}$$
$$\textbf{(red)} \qquad\qquad\qquad\qquad \textbf{(blue)}$$

It is necessary to determine the volume of EDTA used to react with the small amount of magnesium ion attached to the indicator since this volume of EDTA is not related to the concentration of calcium and magnesium ions in the original hard water sample. This is done by titrating a sample of deionized water to which has been added the same amount of Mg^{2+} and Eriochrome Black T that were used in the hard water titration. This is known as **titrating a blank.** *The volume of EDTA required to titrate the blank is subtracted from the volume of EDTA used to reach the endpoint in the hard water titration.* The result is the volume of EDTA required to reach the equivalence point in the hard water titration.

Procedure

This experiment is done in pairs. Carefully clamp a 50 mL **buret** onto a ring stand using a **buret clamp**. Clean the buret by pouring tap water (contained in a beaker) into the top of the buret using a funnel, and open the buret valve so that the tap water can run through the buret and empty into a spare beaker. Next, clean the buret two more times by pouring deionized (DI) water from a clean beaker into the buret through the funnel to rinse out any residual tap water and impurities. With some DI water still in the buret, close the buret valve, remove the buret from the buret clamp, and manually rotate the buret to ensure the inner surface is coated with DI water. Be certain the DI rinse water is allowed to drain from the tip of the buret and from the top. Never place the buret under the tap or DI water spigot to prevent breaking this expensive glassware.

Obtain 40 mL of EDTA solution (approximately 0.010 M) in a labeled 250 mL beaker. Reconnect the buret to the ring stand and then rinse the buret with two 10 mL portions of EDTA in a similar manner to the DI rinse, letting the excess EDTA run out through the buret into a spare beaker. Close the buret valve and fill the buret with the EDTA solution using a funnel. Discharge any air bubbles trapped in the tip of the buret.

Part A: Titration of Blank

Using a 25.00 mL **volumetric pipet,** place 25.00 mL of deionized water in a 250 mL Erlenmeyer flask. A volumetric pipet is a unique piece of glassware designed to hold and deliver a precalibrated volume of liquid and is much more accurate for measuring volume than a graduated cylinder. Allow the tip of the pipet to touch the side of the flask and let the sample solution flow out by gravity. Now, add 5 mL of pH 10 buffer, 2 drops of Eriochrome Black T indicator, and 15 drops of 0.03 M $MgCl_2$ to the flask. The solution in the flask should be red.

Titrate this solution by swirling the sample and adding EDTA titrant drop-wise to the flask. Do not leave the funnel in the buret while titrating. As the solution nears the endpoint, a purple color will develop. Continue to swirl the flask and add titrant until the solution turns blue with no tinge of red. Place the flask on a piece of white paper to help you discern color changes. Read the final buret reading to **0.01 mL** and record the data in column Trial 1 on your data sheet. Dispose of the solution and rinse the flask well with DI water for Trial 2. Allow your partner to repeat the titration on a fresh blank and record the data under the column labeled Trial 2. A third blank titration does not need to be performed if the EDTA volumes from Trials 1 and 2 are within 0.50 mL.

Save the final titrated blank to use as a color reference during the titration of the hard water sample. The volume of EDTA used in this titration is the amount consumed by the indicator. Average the two blank titrations and record this value as the average volume of EDTA required to titrate the blank.

Part B: Titration of Hard Water Sample

Obtain approximately 100 mL of an unknown hard water sample in a labeled 250 mL beaker. Using a 25 mL volumetric pipet, deliver two 25.00 mL **aliquots** (part of a whole) of the water sample to a 250 mL flask for a total of 50.00 mL. Now add 5 mL of pH 10 buffer, 2 drops of Eriochrome Black T, and 15 drops of 0.03 M $MgCl_2$. Titrate the contents of the flask with EDTA to a blue endpoint. Allow your partner to repeat the procedure on a new 50 mL aliquot of the hard water for Trial 2. If the volumes of EDTA used in the two titrations do not agree within 0.50 mL, it is necessary to perform a third trial. Subtract the average volume of EDTA required for the blank titration from the total volume of EDTA used to titrate the hard water sample. This corrects for the amount of EDTA that reacted with the indicator.

Calculations

The endpoint of the titration indicates when the moles of EDTA added from the buret equals the total moles of calcium and magnesium ions (reported as $CaCO_3$) in the sample. The following relationship exists **at the equivalence point** of the titration.

moles of EDTA = moles of CaCO$_3$

1. To determine the moles of calcium in the sample (as $CaCO_3$):

$$L_{EDTA} \times M_{EDTA} \times \frac{(1 \text{ mole CaCO}_3)}{(1 \text{ mole EDTA})} = \text{moles CaCO}_3 \text{ in sample}$$

2. To convert moles to molarity of $CaCO_3$:

$$\frac{\text{moles CaCO}_3}{50.00 \times 10^{-3} \text{ L sample}} = \text{M CaCO}_3$$

3. To convert molarity to mg/L of $CaCO_3$:

$$\text{M CaCO}_3 \times \frac{1.00 \times 10^2 \text{ g CaCO}_3}{\text{mole}} \times \frac{(1000 \text{ mg})}{\text{g}} = \text{mg/L CaCO}_3$$

4. Units of parts per million (ppm) are often used when discussing small quantities of material. ppm refers to the number of grams of $CaCO_3$ present in one million grams of water analyzed. Water hardness is reported as ppm $CaCO_3$ which is the same as mg/L $CaCO_3$ in water solutions: **ppm = mg / L**

Safety Precautions

The Eriochrome Black T dye may cause stains in clothing, therefore be cautious when using this indicator. The pH 10 buffer contains ammonium hydroxide and will be located in the fume hood. Add the buffer to your sample in your flask will dilute the buffer and help you avoid breathing any vapors from the buffer.

Waste Management

Waste can be rinsed down the drain with plenty of water.

Water Hardness: Determination with EDTA Prelab Questions

Name _____ **Section** _____

1. Explain the reason for doing the blank titration. Explain how this volume affects the titration of a water sample with EDTA.

2. a) If a student has 4.6×10^{-3} moles of Ca^{2+} in a tap water sample, how many moles of EDTA must be added to reach the equivalence point?

   b) *If* the stoichiometry of the above reaction was not 1:1, but 3 moles of Ca^{2+} reacted for every 1 mole of EDTA available, how many moles of EDTA would be required at the equivalence point using the same tap water sample?

Water Hardness: Determination with EDTA Data Sheet

Name _____ Section _____

Name of Lab Partner(s) _____

Molarity of EDTA _____

Part A: Blank Titration

	Trial 1	Trial 2	Trial 3
Final buret reading			
Initial buret reading			
Volume of EDTA to titrate blank			

Average volume of EDTA required to titrate blank _____

Part B: Titration of Unknown Hard Water Sample
Unknown Number _____

	Trial 1	Trial 2	Trial 3
Final buret reading			
Initial buret reading			
Total Volume of EDTA to titrate sample			
Average volume of EDTA to titrate blank (Part A)			
Corrected Volume of EDTA to titrate unknown hard water sample			

Average volume of EDTA required to titrate hard water _____

Molarity of $CaCO_3$ in hard water _____

ppm of $CaCO_3$ in hard water _____

Water Hardness: Determination with EDTA Review Questions

Name _____ **Section** _____

1. A water sample is found to have a Cl^- content of 100 ppm as NaCl. What is the concentration of chloride in moles per liter?

2. What is the molarity of $CaCO_3$ in a hard water sample if a 25.0 mL sample requires 12.54 mL of 0.0100 M EDTA to reach the end point? The volume for the blank titration was 2.58 mL.

3. A 100 mL sample of hard water required 36.85 mL of 0.0200 M EDTA to reach the endpoint in the titration. The blank titration volume was 3.20 mL. Determine the hardness of the water.

Water Hardness: Determination with EDTA Supplemental Questions

Name _____ **Section** _____

Appendix A
Material Safety Data Sheet

Ethanol, Absolute

ACC# 89308

Section 1—Chemical Product and Company Identification

MSDS Name: Ethanol, Absolute

Catalog Numbers: NC9602322

Synonyms: Ethyl Alcohol; Ethyl Alcohol Anhydrous; Ethyl Hydrate; Ethyl Hydroxide; Fermentation Alcohol; Grain Alcohol; Methylcarbinol; Molasses Alcohol; Spirits of Wine.

Company Identification:

Fisher Scientific

1 Reagent Lane

Fair Lawn, NJ 07410

For information, call: 201-796-7100

Emergency Number: 201-796-7100

For CHEMTREC assistance, call: 800-424-9300

For International CHEMTREC assistance, call: 703-527-3887

Section 2—Composition, Information on Ingredients

CAS#	Chemical Name	Percent	EINECS/ELINCS
64-17-5	Ethanol	ca.100	200-578-6

Hazard Symbols: F

Risk Phrases: 11

Section 3—Hazards Identification

Emergency Overview

Appearance: colorless clear liquid. Flash Point: 16.6 deg C. **Flammable liquid and vapor.** May cause central nervous system depression. Causes severe eye irritation. Causes respiratory tract irritation. Causes moderate skin irritation. This substance has caused adverse reproductive and fetal effects in humans. **Warning!** May cause liver, kidney and heart damage.

Reprinted by permission of Fisher Scientific.

Target Organs: Kidneys, heart, central nervous system, liver.

Potential Health Effects

Eye: Causes severe eye irritation. May cause painful sensitization to light. May cause chemical conjunctivitis and corneal damage.

Skin: Causes moderate skin irritation. May cause cyanosis of the extremities.

Ingestion: May cause gastrointestinal irritation with nausea, vomiting and diarrhea. May cause systemic toxicity with acidosis. May cause central nervous system depression, characterized by excitement, followed by headache, dizziness, drowsiness, and nausea. Advanced stages may cause collapse, unconsciousness, coma and possible death due to respiratory failure.

Inhalation: Inhalation of high concentrations may cause central nervous system effects characterized by nausea, headache, dizziness, unconsciousness and coma. Causes respiratory tract irritation. May cause narcotic effects in high concentration. Vapors may cause dizziness or suffocation.

Chronic: May cause reproductive and fetal effects. Laboratory experiments have resulted in mutagenic effects. Animal studies have reported the development of tumors. Prolonged exposure may cause liver, kidney, and heart damage.

Section 4—First Aid Measures

Eyes: Get medical aid. Gently lift eyelids and flush continuously with water.

Skin: Get medical aid. Wash clothing before reuse. Flush skin with plenty of soap and water.

Ingestion: Do NOT induce vomiting. If victim is conscious and alert, give 2–4 cupfuls of milk or water. Never give anything by mouth to an unconscious person. Get medical aid.

Inhalation: Remove from exposure and move to fresh air immediately. If not breathing, give artificial respiration. If breathing is difficult, give oxygen. Get medical aid. Do NOT use mouth-to-mouth resuscitation.

Notes to Physician: Treat symptomatically and supportively. Persons with skin or eye disorders or liver, kidney, chronic respiratory diseases, or central and peripheral nervous sytem diseases may be at increased risk from exposure to this substance.

Antidote: None reported.

Section 5—Fire Fighting Measures

General Information: Containers can build up pressure if exposed to heat and/or fire. As in any fire, wear a self-contained breathing apparatus in pressure-demand, MSHA/NIOSH (approved or equivalent), and full protective gear. Vapors may form an explosive mixture with air. Vapors can travel to a source of ignition and flash back. Will burn if involved in a fire. Flammable Liquid. Can release vapors that form explosive mixtures at temperatures above the flashpoint. Use water spray to keep fire-exposed containers cool. Containers may explode in the heat of a fire.

Extinguishing Media: For small fires, use dry chemical, carbon dioxide, water spray or alcohol-resistant foam. For large fires, use water spray, fog, or alcohol-resistant foam. Use water spray to cool fire-exposed containers. Water may be ineffective. Do NOT use straight streams of water.

Flash Point: 16.6 deg C (61.88 deg F)

Autoignition Temperature: 363 deg C (685.40 deg F)

Explosion Limits, Lower: 3.3 vol %

Upper: 19.0 vol %

NFPA Rating: (estimated) Health: 2; Flammability: 3; Instability: 0

Section 6—Accidental Release Measures

General Information: Use proper personal protective equipment as indicated in Section 8.

Spills/Leaks: Absorb spill with inert material (e.g. vermiculite, sand or earth), then place in suitable container. Remove all sources of ignition. Use a spark-proof tool. Provide ventilation. A vapor suppressing foam may be used to reduce vapors.

Section 7—Handling and Storage

Handling: Wash thoroughly after handling. Use only in a well-ventilated area. Ground and bond containers when transferring material. Use spark-proof tools and explosion proof equipment. Avoid contact with eyes, skin, and clothing. Empty containers retain product residue, (liquid and/or vapor), and can be dangerous. Keep container tightly closed. Avoid contact with heat, sparks and flame. Avoid ingestion and inhalation. Do not pressurize, cut, weld, braze, solder, drill, grind, or expose empty containers to heat, sparks or open flames.

Storage: Keep away from heat, sparks, and flame. Keep away from sources of ignition. Store in a tightly closed container. Keep from contact with oxidizing materials. Store in a cool, dry, well-ventilated area away from incompatible substances. Flammables-area. Do not store near perchlorates, peroxides, chromic acid or nitric acid.

Section 8—Exposure Controls, Personal Protection

Engineering Controls: Use explosion-proof ventilation equipment. Facilities storing or utilizing this material should be equipped with an eyewash facility and a safety shower. Use adequate general or local exhaust ventilation to keep airborne concentrations below the permissible exposure limits.

Exposure Limits

Chemical Name	ACGIH	NIOSH	OSHA - Final PELs
Ethanol	1000 ppm TWA	1000 ppm TWA; 1900 mg/m3 TWA 3300 ppm IDLH	1000 ppm TWA; 1900 mg/m3 TWA

OSHA Vacated PELs: Ethanol: 1000 ppm TWA; 1900 mg/m3 TWA

Personal Protective Equipment

Eyes: Wear appropriate protective eyeglasses or chemical safety goggles as described by OSHA's eye and face protection regulations in 29 CFR 1910.133 or European Standard EN166.

Skin: Wear appropriate protective gloves to prevent skin exposure.

Clothing: Wear appropriate protective clothing to prevent skin exposure.

Respirators: A respiratory protection program that meets OSHA's 29 CFR 1910.134 and ANSI Z88.2 requirements or European Standard EN 149 must be followed whenever workplace conditions warrant a respirator's use.

Section 9—Physical and Chemical Properties

Physical State: Clear liquid

Appearance: colorless

Odor: Mild, rather pleasant, like wine or whis

pH: Not available.

Vapor Pressure: 59.3 mm Hg @ 20 deg C

Vapor Density: 1.59

Evaporation Rate: Not available.

Viscosity: 1.200 cP @ 20 deg C

Boiling Point: 78 deg C

Freezing/Melting Point: −114.1 deg C

Decomposition Temperature: Not available.

Solubility: Miscible.

Specific Gravity/Density: 0.790 @ 20°C

Molecular Formula: C2H5OH

Molecular Weight: 46.0414

Section 10—Stability and Reactivity

Chemical Stability: Stable under normal temperatures and pressures.

Conditions to Avoid: Incompatible materials, ignition sources, excess heat, oxidizers.

Incompatibilities with Other Materials: Strong oxidizing agents, acids, alkali metals, ammonia, hydrazine, peroxides, sodium, acid anhydrides, calcium hypochlorite, chromyl chloride, nitrosyl perchlorate, bromine pentafluoride, perchloric acid, silver nitrate, mercuric nitrate, potassium-tert-butoxide, magnesium perchlorate, acid chlorides, platinum, uranium hexafluoride, silver oxide, iodine heptafluoride, acetyl bromide, disulfuryl difluoride, tetrachlorosilane + water, acetyl chloride, permanganic acid, ruthenium (VIII) oxide, uranyl perchlorate, potassium dioxide.

Hazardous Decomposition Products: Carbon monoxide, irritating and toxic fumes and gases, carbon dioxide.

Hazardous Polymerization: Will not occur.

Section 11—Toxicological Information

RTECS#:

CAS# 64-17-5: KQ6300000

LD50/LC50:

CAS# 64-17-5:

Draize test, rabbit, eye: 500 mg Severe;

Draize test, rabbit, eye: 500 mg/24H Mild;

Draize test, rabbit, skin: 20 mg/24H Moderate;

Inhalation, mouse: LC50 = 39 gm/m3/4H;

Inhalation, rat: LC50 = 20000 ppm/10H;

Oral, mouse: LD50 = 3450 mg/kg;

Oral, rabbit: LD50 = 6300 mg/kg;

Oral, rat: LD50 = 9000 mg/kg;

Oral, rat: LD50 = 7060 mg/kg;<BR.

Carcinogenicity:

CAS# 64-17-5:

ACGIH: A4 - Not Classifiable as a Human Carcinogen

Epidemiology: Ethanol has been shown to produce fetotoxicity in the embry o or fetus of laboratory animals. Prenatal exposure to ethanol is associated with a distinct pattern of congenital malformations that have collecetively been termed the "fetal alcohol syndrome".

Teratogenicity: Oral, Human - woman: TDLo = 41 gm/kg (female 41 week(s) after conception) Effects on Newborn - Apgar score (human only) and Effects on Newborn - other neonatal measures or effects and Effects on Newborn - drug dependence.

Reproductive Effects: Intrauterine, Human - woman: TDLo = 200 mg/kg (female 5 day(s) pre-mating) Fertility - female fertility index (e.g. # females pregnant per # sperm positive females; # females pregnant per # females mated).

Neurotoxicity: No information available.

Mutagenicity: DNA Inhibition: Human, Lymphocyte = 220 mmol/L.; Cytogenetic Analysis: Human, Lymphocyte = 1160 gm/L.; Cytogenetic Analysis: Human, Fibroblast = 12000 ppm.; Cytogenetic Analysis: Human, Leukocyte = 1 pph/72H (Continuous).; Sister Chromatid Exchange: Human, Lymphocyte = 500 ppm/72H (Continuous).

Other Studies: Standard Draize Test(Skin, rabbit) = 20 mg/24H (Moderate) Standard Draize Test: Administration into the eye (rabbit) = 500 mg (Severe).

Section 12—Ecological Information

Ecotoxicity: Fish: Rainbow trout: LC50 = 12900–15300 mg/L; 96 Hr; Flow-through @ 24–24.3°C Rainbow trout: LC50 = 11200 mg/L; 24 Hr; Fingerling (Unspecified) ria: Phytobacterium phosphoreum: EC50 = 34900 mg/L; 5–30 min; Microtox test When spilled on land it is apt to volatilize, biodegrade, and leach into the ground water, but no data on the rates of these processes could be found. Its fate in ground water is unknown. When released into water it will volatilize and probably biodegrade. It would not be expected to adsorb to sediment or bioconcentrate in fish.

Environmental: When released to the atmosphere it will photodegrade in hours (polluted urban atmosphere) to an estimated range of 4 to 6 days in less polluted areas. Rainout should be significant.

Physical: No information available.

Other: No information available.

Section 13—Disposal Considerations

Chemical waste generators must determine whether a discarded chemical is classified as a hazardous waste. US EPA guidelines for the classification determination are listed in 40 CFR Parts 261.3. Additionally, waste generators must consult state and local hazardous waste regulations to ensure complete and accurate classification.

RCRA P-Series: None listed.

RCRA U-Series: None listed.

Section 14—Transport Information

	US DOT	IATA	RID/ADR	IMO	Canada TDG
Shipping Name:	No information available.				No information available.
Hazard Class:					
UN Number:					
Packing Group:					

Section 15—Regulatory Information

US Federal

TSCA

CAS# 64-17-5 is listed on the TSCA inventory.

Health & Safety Reporting List

None of the chemicals are on the Health & Safety Reporting List.

Chemical Test Rules

None of the chemicals in this product are under a Chemical Test Rule.

Section 12b

None of the chemicals are listed under TSCA Section 12b.

TSCA Significant New Use Rule

None of the chemicals in this material have a SNUR under TSCA.

SARA

CERCLA Hazardous Substances and corresponding RQs

None of the chemicals in this material have an RQ.

SARA Section 302 Extremely Hazardous Substances

None of the chemicals in this product have a TPQ.

SARA Codes

CAS # 64-17-5: acute, chronic, flammable.

Section 313

No chemicals are reportable under Section 313.

Clean Air Act:

This material does not contain any hazardous air pollutants. This material does not contain any Class 1 Ozone depletors. This material does not contain any Class 2 Ozone depletors.

Clean Water Act:

None of the chemicals in this product are listed as Hazardous Substances under the CWA. None of the chemicals in this product are listed as Priority Pollutants under the CWA. None of the chemicals in this product are listed as Toxic Pollutants under the CWA.

OSHA:

None of the chemicals in this product are considered highly hazardous by OSHA.

STATE

CAS# 64-17-5 can be found on the following state right to know lists: California, New Jersey, Pennsylvania, Minnesota, Massachusetts.

WARNING: This product contains Ethanol, a chemical known to the state of California to cause birth defects or other reproductive harm. California No Significant Risk Level: None of the chemicals in this product are listed.

European/International Regulations

European Labeling in Accordance with EC Directives

Hazard Symbols:

F

Risk Phrases:

R 11 Highly flammable.

Safety Phrases:

S 16 Keep away from sources of ignition - No

smoking.

S 33 Take precautionary measures against static

discharges.

S 7 Keep container tightly closed.

S 9 Keep container in a well-ventilated place.

WGK (Water Danger/Protection)

CAS# 64-17-5: 0

Canada - DSL/NDSL

CAS# 64-17-5 is listed on Canada's DSL List.

Canada - WHMIS

This product has a WHMIS classification of B2, D2A.

Canadian Ingredient Disclosure List

CAS# 64-17-5 is listed on the Canadian Ingredient Disclosure List.

Exposure Limits

CAS# 64-17-5: OEL-AUSTRALIA:TWA 1000 ppm (1900 mg/m3) OEL-BELGIUM:TWA 1000 ppm (1880 mg/m3) OEL-CZECHOSLOVAKIA:TWA 1000 mg/m3;STEL 5000 mg/m3 OEL-DENMARK:TWA

1000 ppm (1900 mg/m3) OEL-FINLAND:TWA 1000 ppm (1900 mg/m3);STEL 1250 ppm (2400 mg/m3) OEL-FRANCE:TWA 1000 ppm (1900 mg/m3);STEL 5000 pp OEL-GERMANY:TWA 1000 ppm (1900 mg/m3) OEL-HUNGARY:TWA 1000 mg/m3;STEL 3000 mg/m3 OEL-THE NETHERLANDS:TWA 1000 ppm (1900 mg/m3) OEL-THE PHILIPPINES:TWA 1000 ppm (1900 mg/m3) OEL-POLAND :TWA 1000 mg/m3 OEL-RUSSIA:STEL 1000 mg/m3 OEL-SWEDEN:TWA 1000 ppm (1900 mg/m3) OEL-SWITZERLAND:TWA 1000 ppm (1900 mg/m3) OEL-THAILAND:TWA 1000 ppm (1900 mg/m3) OEL-TURKEY:TWA 1000 ppm (1900 mg/m3) OEL-UNITED KINGDOM:TWA 1000 ppm (1900 mg/m3) JAN9 OEL IN BULGARIA, COLOMBIA, JORDAN, KOREA check ACGIH TLV OEL IN NEW ZEALAND, SINGAPORE, VIETNAM check ACGI TLV

Section 16—Additional Information

MSDS Creation Date: 7/27/1999

Revision #4 Date: 3/18/2003

The information above is believed to be accurate and represents the best information currently available to us. However, we make no warranty of merchantability or any other warranty, express or implied, with respect to such information, and we assume no liability resulting from its use. Users should make their own investigations to determine the suitability of the information for their particular purposes. In no event shall Fisher be liable for any claims, losses, or damages of any third party or for lost profits or any special, indirect, incidental, consequential or exemplary damages, howsoever arising, even if Fisher has been advised of the possibility of such damages.

Appendix B
Summary VSEPR and Hybridization Table

Electron Domains	Electron-Domain Geometry	Predicted Bond Angle(s)	Hybridization of Central Atom	Molecular Geometry		
				0 Lone Pair	1 Lone Pair	2 Lone Pair
2	Linear	180°	sp	Linear		
3	Trigonal Planar	120°	sp^2	Trigonal Planar	Bent	
4	Tetrahedral	109.5°	sp^3	Tetrahedral	Trigonal Pyramidal	Bent
5	Trigonal Bipyramidal	90°, 120°	sp^3d	Trigonal Bipyramidal	Seesaw	T-shaped
6	Octahedral	90°	sp^3d^2	Octahedral	Square Pyramidal	Square Planar

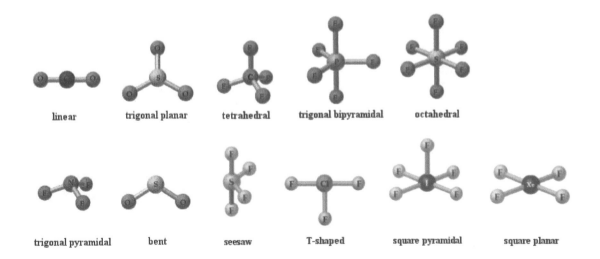

linear   trigonal planar   tetrahedral   trigonal bipyramidal   octahedral

trigonal pyramidal   bent   seesaw   T-shaped   square pyramidal   square planar

Appendix C
Set-up of MicroLab Equipment

Introduction

The MicroLab Data Acquisition Equipment used in the General Chemistry 101/102 Laboratory is a data processing unit designed to convert electrical signals from one or more sensors and produce digital readouts through an interfacing computer. The computer software can provide this data through real-time readouts, in table form, and through graph relationships. The MicroLab Equipment will be used for five main types of experiments in the 101/102 curriculum: time vs. temperature determinations, gas laws (pressure), pH determinations, spectrophotometry, and electrochemistry.

Procedure

Connect the MicroLab unit to the designated USB port on the lab computer. Turn on the power to the MicroLab unit by pushing in the green button on the upper right hand corner of the unit. Ask your instructor to install the MicroLab software on the lab computer if necessary. Click on the MicroLab folder on the Desktop and wait until the program loads. Once the program has loaded, click on the "Time and Temperature" prompt, as in the photo below:

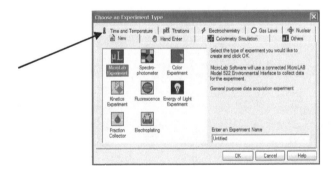

If the drop down menu appears with choices for CHML 101 and 102 experiments, you may open the appropriate program and begin the day's experiment.

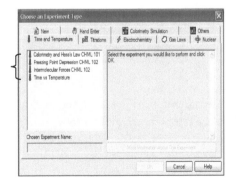

If a menu with CHML 101/102 experiments does not appear, then you must go through the sequence in the following slides to create parameters for your experiment. An instructional video can also be found on the MicroLab website: http://www.microlabinfo.com/Instructional_videos.html

Developing an Experimental Program

Double click on the general program labeled "Time and Temperature", or on a specific experiment program from the drop down menu. When the program is opened, you will need to remove the sensors by highlighting one at a time and clicking "remove" as in the photo below. A prompt box will appear to warn you that removing the sensor will remove all data with the sensor—click "okay."

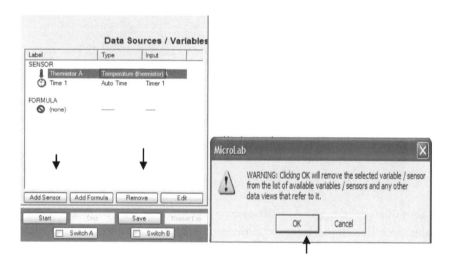

Click the "Add sensor" menu tab. A drop down menu of various choices of sensors will appear. If performing a "Temperature vs. Time" experiment you will need to select the "Temperature (thermistor)" sensor. This sensor will now appear in the main Data Sources/Variables box of your experimental program.

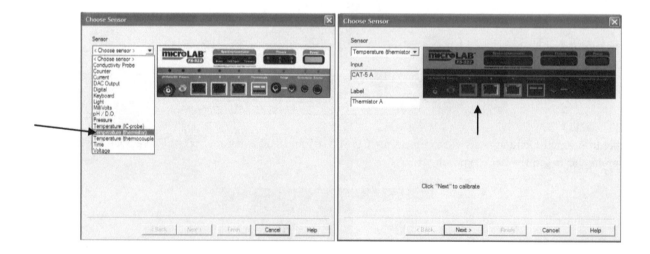

Once the "Temperature (thermistor)" sensor has been selected you will be asked to select the Input. Move the cursor to the picture of Port A to select the temperature probe Input, as in the photo above. You may connect your temperature probe to Port A at this time. Click "Next."

Calibrating Sensors

Now you will be asked to calibrate the sensor you have chosen. Click on the menu item "Read Calibration From a File." Click on "Model 103 Thermistor.cft" to indicate the type of temperature probe you are using. Click "Finish."

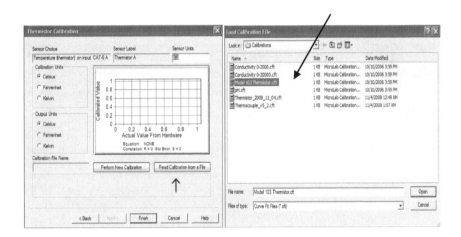

Remove the Timing Sensor from the Data Sources menu if the timing function is not active. Now you can add the Timing Sensor to your experimental template. Click on the menu item "Add Sensor." Select the "Time" sensor from the drop down menu of sensor choices.

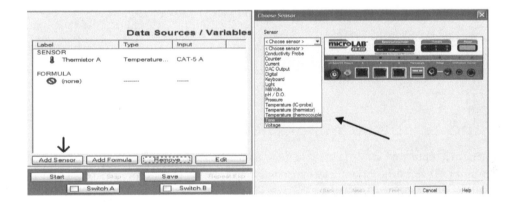

Click the appropriate Port 1 for Timing Input as shown in the photo at the right. Click "Next" to set the calibration data for the timing sensor. Click "Finish" to accept the default settings for the internal timing sensor.

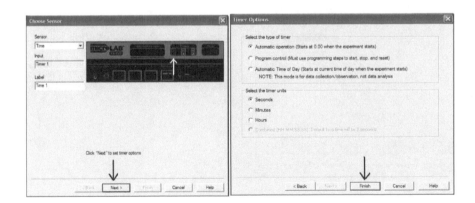

Designations for Graph, Table, and Real Time Visuals

To visualize the sensor data collected along a graph axis, in table form, and as real-time information, you must click and drag the sensor to the appropriate locations. Click and drag one sensor at a time to the appropriate axis of the graph, then to the appropriate column of the table, and finally, to the Digital Display box.

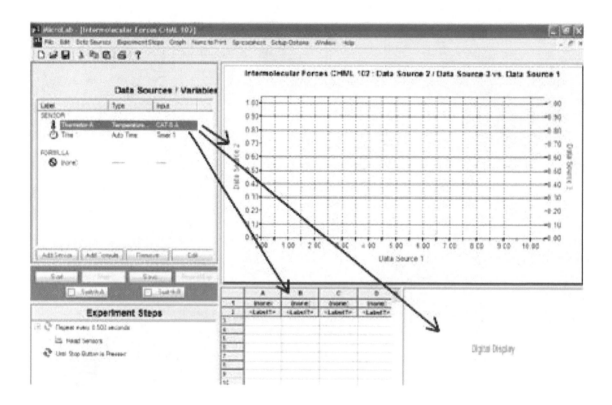

Naming and Saving an Experimental Template

To save experiment as a template for future use you must name the file by clicking on "File" and choosing the "Save As Template" prompt. Type in the title of the experiment and click the appropriate experiment type. (Time and Temperature is shown below for the Intermolecular Forces Exp).

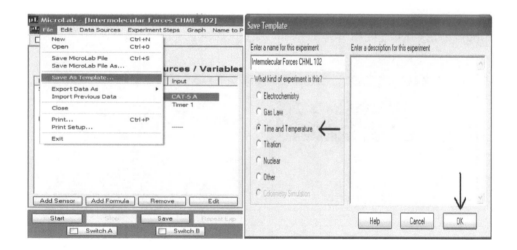

Experimental Prompts

To initiate an experiment, click the "Start" button. To stop an experiment, click the "Stop" button.

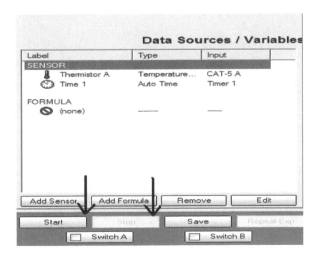

To begin collecting new data using the same experimental parameters, click the "Repeat Exp" button. In the prompt box which follows, ALWAYS click "Repeat experiment without saving data," as in the photo below.

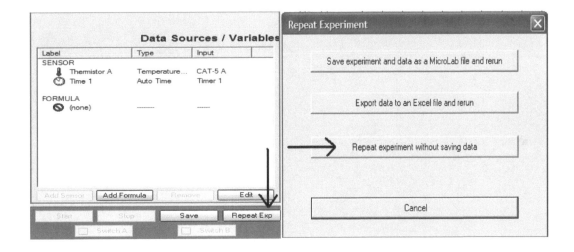

Appendix D
Spectrophotometry

Introduction

Colored solutions, when irradiated with white light, selectively absorb certain wavelengths but not others. As the solution becomes more concentrated, the amount of light absorbed increases and its color intensifies. Once the wavelength of maximum absorption for a compound is known, we can use this behavior to determine the concentration of the compound in a solution.

In a **UV-Vis Spectrophotometer,** a monochromatic (single wavelength) light beam passes through a solution and strikes a detector, which registers the intensity of the light. By comparing the intensity of light passing through the sample with the intensity of light passing through a blank (a solution that does not contain the light absorbing compound), the solution's **absorbance** can be determined. If only one substance in the colored solution absorbs light, then the absorbance of the solution is directly proportional to both the concentration of the substance and the width of the tube holding the sample. The relationship between the concentration of a light absorbing compound in a solution and how much light the solution absorbs is know as the **Beer-Lambert Law**.

$$A = abc$$

In this equation, **A** stands for absorbance, **b** is the pathlength of light through the sample, **c** is the concentration (molarity) of the solution, and **a** is called the molar absorptivity of the solute. The molar absorptivity is a constant for a given solute at a particular wavelength. Most experiments are designed in such a way that the values of a and b are held constant. This means that our only variables are **concentration** and **absorbance.** Therefore, by measuring the absorbance of a colored solution, we can determine the concentration of the light absorbing substance in the solution.

The Beer-Lambert Law indicates that the relationship between the absorbance of a colored solution and the concentration of light absorbing compound in the solution is linear. By determining the amount of light absorbed by a series of solutions of know concentration, we can construct a **calibration curve** that allows us to determine the concentration of any solution that contains the same compound by simply measuring its absorbance. This is illustrated in Figure 1.

In General Chemistry, you will measure the absorbance of colored solutions using a spectrophotometer that is part of the MicroLab data acquisition system, sometimes called the MicroLab Spectrophotometer. This simple spectrophotometer contains a series of 10–16 colored LEDs that cover the wavelength range from 400 to 644 nm, or 390 to 940 nm in newer MicroLab units. Measuring concentrations using a spectrophotometer is quick and it does not disturb the chemical composition of the system being studied.

General Instructions for Making Absorbance Measurements using the MicroLab Data Acquisition Equipment

Clean and dry several **cuvettes** (glassware made specifically for the spectrophotometer of uniform thickness and clarity). Do not clean cuvettes with test tube wire brushes, the scratches will interfere with your absorbance measurements. Place enough solution in the cuvette so that it is filled about 3/4-full. Do not allow the solution

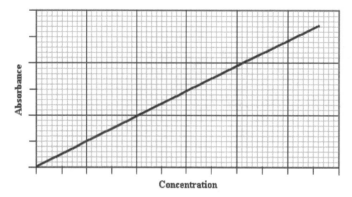

FIGURE 1 — *Calibration Curve of Absorbance vs. Concentration*

to spill into the instrument. Before placing a cuvette in the instrument, wipe the outside of the cuvette with a Kimwipe to remove fingerprints.

Measuring Absorbance

Setup your laboratory computer and the MicroLab data acquisition system. Start the MicroLab software and select Spectrophotometer Experiment. The following instructions assume that you have already prepared a **blank** (a solution that contains everything that is used to prepare the standards but does <u>not</u> contain the analyte of interest), a set of **standards** (solutions containing known concentrations of the analyte of interest), and your unknown sample(s).

1. Place the blank in the spectrophotometer and press the "Read Blank" button.

2. Click on "Absorbance" at the top of the screen. This will switch the spectrophotometer readings from % transmittance to absorbance.

3. Remove the blank and place the standard with the lowest concentration of analyte in the spectrophotometer. Press the "Add" button. In the dialog box that appears, enter "Std 1" for the Sample ID and enter the actual concentration of the standard in the Concentration box. Press the "OK" button.

4. Click on the colored bar that represents the desired wavelength for the analysis.

5. Repeat step (3) for the remaining standards labeling them "Std 2", "Std 3", etc.

6. Click on the tab labeled "3 Curve" and select "linear" for the curve fit. The computer will automatically generate your calibration curve.

7. Place the cuvette that contains your first unknown sample in the spectrophotometer. Click on the tab labeled "4 Read". Press the "Add" button and enter "Sample 1" for the Sample ID. Press the "OK" button. The computer will now display the absorbance and the equilibrium concentration of the analyte.

8. Repeat step (7) for any remaining unknown samples.

Figure 2 below is a screen shot of an analysis performed with the MicroLab spectrophotometer using a blank, two standards, and one unknown sample.

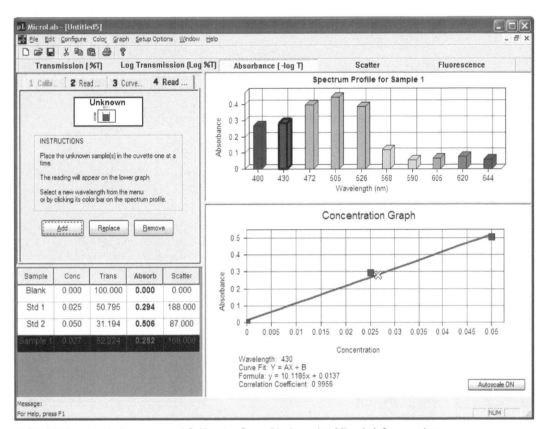

FIGURE 2 — *Profile Spectrum and Calibration Curve Display using MicroLab Spectrophotometer*

Appendix E
Proper Care of pH Electrodes

General Instructions

1. Fill a small plastic wash bottle with deionized (DI) water. This will be used to wash the pH electrode.

2. Fill a beaker with DI water to hold the pH electrode when you are not taking measurements. The electrode must remain wet at all times in order to work properly.

3. Remove the pH electrode from its supplied pH 7 buffer storage solution. Do not return the electrode to this buffer solution until the experiment has been completed to avoid contaminating the storage solution.

4. Wash the electrode with DI water from the wash bottle. Place the electrode in the beaker of DI water.

5. When you are ready to use the pH electrode with your sample, rinse off the electrode with the wash bottle and gently wipe off any residual water with a Kimwipe.

6. Place the probe in the sample solution. When taking a reading, make certain the solution covers the small hole on the side of the electrode located above the glass bulb. It may be necessary to add more solution or you may tilt the beaker so that the hole is covered. After recording the pH reading, remove the electrode from the solution, rinse it thoroughly with DI water, and return it to the storage beaker filled with DI water.

7. Repeat these steps for subsequent sample measurements.

8. When all readings have been taken, wash the electrode completely, dry it gently with a Kimwipe and return it to the pH 7 buffer storage solution.

Titrations

When using the pH meter to perform a potentiometric titration, the electrode will remain in the solution throughout the titration as the titrant is added. You may use the electrode to gently stir the solution to ensure that equilibrium has been established. Upon completion of the titration, the electrode should be rinsed thoroughly with DI water, dried gently with a Kimwipe, and returned to the pH 7 storage solution.

Appendix F
Supplemental Questions

1. Describe the following nanoscale view of a sample of matter using terms such as pure, mixture, element, compound, solid, liquid, gas, heterogeneous, and homogeneous.

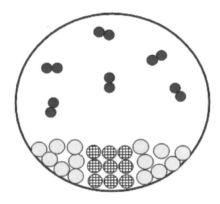

2. Indicate which of the following processes are physical changes or chemical changes and explain the reason:
   a) Corrosion of iron metal
   b) Evaporation of water
   c) Pulverizing an aspirin

3. Label each of the following as either a physical process or a chemical process:
   a) Tearing a sheet of paper
   b) Boiling water
   c) Frying an egg

4. Indicate which of the following are exact numbers:
   a) The mass of a paper clip
   b) The number of inches in a mile
   c) The number of nanometers in a centimeter

5. A reddish-colored metal is placed in a burner flame where it turns black and gains mass. Does this represent an element forming a compound (or compounds) or a compound which forms elements?

6. A solid blue substance A is heated in air. As it is heated, it gives off a gas B and the original solid blue substance turns into a white solid, C. The gas B has the same properties as the product formed by the direct reaction of oxygen and hydrogen. Can you describe substances A, B, and C as elements or compounds?

7. Answer the following questions using figures a) to i) shown below. (Each question may have more than one answer.)

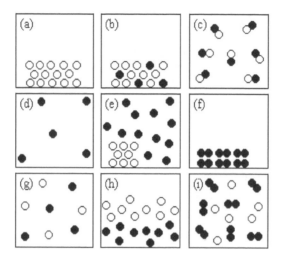

a) Which represent nanoscale particles in a sample containing *only* a solid?
b) Which represent nanoscale particles in a sample containing *only* a liquid?
c) Which represent nanoscale particles in a sample containing *only* a gas?
d) Which represent nanoscale particles in a sample containing *only* an element?
e) Which represent nanoscale particles in a sample containing *only* a compound?
f) Which represent nanoscale particles in a sample containing *only* a pure substance?
g) Which represent nanoscale particles in a sample of a mixture?

8. Indicate the number of significant figures in each of the following measured quantities:
    a) 1.09×10^{-3} km
    b) 0.0234 m^2
    c) 7,300 cm

9. A Morgan silver dollar has a mass of 26.73 g. By law, it must contain 90% silver. In today's market, silver sells for $5.30 per troy ounce (1 troy ounce = 31.1 g). What is the value of the silver in a Morgan silver dollar?

10. Winds are forecast for 15 miles per hour. What is the speed expressed in meters per second?
    (1 km = 0.621 miles)

11. If you have a ring that contains 1.94 g of gold, how many atoms of gold are in the ring?

12. What is the density (expressed in g/cm^3) of a 1.000 kg block that is 3.00 feet × 2.00 inches × 6.00 inches?

13. The density of copper is 9.0 g/cm^3. What is the mass of a 12 cm length of copper wire with a diameter of 0.15 cm? ($V_{cylinder} = \pi r^2 h$).

14. Three different analyst measure the density of a sample. Each analyst determined the density three times and then reported the average and average deviation of the density. The results are as follows: Analyst A: 5.65 ± 0.03 g/mL, Analyst B: 5.44 ± 0.05 g/mL, Analyst C: 5.78 ± 0.07 g/mL. The actual density of the item is 5.74 g/mL.
    a) Rank the analysts from worst to best for accuracy of their results:
       Accuracy: _____ _____ _____
                        worst   best

b) Rank the analysts from worst to best for precision of their results:

Precision: _____ _____ _____

               worst   best

15. The density of mercury is 13.6 g/mL. What is its density in SI units?

16. The mass of a 100 g standard weight from the National Institute of Standards and Technology is 100.000 g. This standard weight is weighed three times on a balance. The masses determined were: 99.996 g, 99.999 g, and 99.997 g. What is the average mass of the standard given by this balance? What is the average deviation?

17. Two students determine the percentage of lead in a sample as a laboratory exercise. The true percentage is known to be 23.65. The students' results for three determinations are shown below:

    Student 1: 23.66, 23.78, 23.47
    Student 2: 23.42, 23.38, 23.40

a) Calculate the average for each set of data. Explain why one set is more accurate than the other.

b) One way to determine the precision of each data set is to calculate what is called the average absolute deviation. To do this, subtract each data point in a set from the average value for the set. This is called the deviation. Make each deviation positive (absolute value) and then calculate the average deviation for each data set. Which data set has the greatest precision?

18. Complete the following table using the information on a periodic table.

Symbol	^{39}K	^{35}Cl	
Protons			16
Neutrons			
Electrons			16
Mass number			32
Type of Element*			

* = metal, nonmetal, or metalloid

19. Using the information on a periodic table fill in the table below. (Assume all atoms are neutral.)

Symbol		^{86}Rb	
Protons	46		
Neutrons	60		118
Electrons			79
Mass number			

20. Potassium has three stable isotopes, ^{39}K, ^{40}K, and ^{41}K.

a) How many neutrons are in the nucleus of each isotope?

b) How many electrons are in a potassium atom?

c) How many electrons are in a potassium ion?

21. There are <u>two</u> different isotopes of bromine atoms. Elemental bromine consists of Br_2 molecules and the mass of a Br_2 molecule is the sum of the masses of the two atoms in the molecule. Mass spectrometry shows the following masses and abundances for Br_2 molecules.

Mass (amu) of Br_2 Molecules	Relative Abundance of Br_2 Molecule
157.836	25.69%
159.834	49.99%
161.832	24.31%

    a) What are the masses of the two isotopes of bromine atoms? (Hint: if we call the two isotopes A and B, then the three molecules in the table are AA, AB, and BB.)

    b) Determine the average molecular mass of a Br_2 molecule.

22. An element has three naturally occurring isotopes with the following masses and abundances.
    a) Calculate the average atomic weight of this element.
    b) Which element is it?

Isotopic Mass	Fractional Abundance
27.977	0.9221
28.976	0.0470
29.974	0.0309

23. There are two isotopes of element X. They are shown below with their corresponding percent natural abundances. What is the average atomic mass of element X?
    ^{20}X 10%
    ^{21}X 90%

24. Sulfur trioxide, SO_3, is made in enormous quantities by combining oxygen and sulfur dioxide. The trioxide is not usually isolated, but is converted to sulfuric acid, H_2SO_4. If you have 1.00 lb (454 g) of sulfur trioxide, how many moles does this represent? How many molecules? How may sulfur atoms? How many oxygen atoms?

25. Give the total number of atoms of each element in one formula unit or molecule of each of the following compounds:
    a) $C_6H_5CHCH_2$
    b) $(NH_4)_2SO_4$

26. Calculate the molar mass of Cu_2O and the % by weight of Cu in Cu_2O.

27. Calculate the molar mass of trimethylamine, $N(CH_3)_3$, and the % by weight of H in this compound. How many moles of $N(CH_3)_3$ are contained in 1 g of $N(CH_3)_3$?

28. Determine the empirical and molecular formula of monosodium glutamate (MSG), a flavor enhancer in certain foods, which contains 35.51% C, 4.77% H, 37.85% O, and 13.60% Na, and has a molar mass of 169 g/mole.

29. Calculate the empirical formula for a compound which is 87.5% N and 12.5% H by mass. Calculate the molecular formula for this compound if the molecular weight is 32.0 g/mole.

30. How many moles are represented by 1.00 g of each of the following?
    a) CH_3OH, methanol
    b) $MgSO_4-7H_2O$, magnesium sulfate heptahydrate (Epsom salt)

31. Epinephrine (adrenaline), a hormone secreted into the bloodstream in times of danger or stress, contains 59.0 % C, 7.1 % H, 26.2 % O, and 7.7 % N by mass; its molecular weight is about 180 amu. Determine the empirical and molecular formula of epinephrine.

32. Many familiar substances have common, unsystematic names. For each of the following, give the correct systematic name.
    a) butter of zinc: $ZnCl_2$
    b) burnt ochre: Fe_2O_3
    c) bleach: $NaOCl$
    d) saltpeter: KNO_3
    e) nitrous oxide, also known as laughing gas: N_2O
    f) alcohol sulfuris: CS_2

33. Give the name or formula for each of the following compounds:
    a) NF_3
    b) $Ca(NO_3)_2$
    c) copper(II) phosphate
    d) magnesium sulfate
    e) diphosphorus tetrafluoride
    f) $Co_2(SO_4)_3$
    g) $Al(OH)_3$

34. How many moles of ethanol are contained in 1.00 g of ethanol, C_2H_5OH?

35. Write balanced chemical equations for the word descriptions below.
    a) Zinc(II) carbonate can be heated to form zinc(II) oxide and carbon dioxide.
    b) On treatment with hydrofluoric acid (HF), silicon dioxide forms silicon tetrafluoride and water.

36. Balance the equation for the combustion of ethanol:
    $$C_2H_6O + O_2 \longrightarrow CO_2 + H_2O$$

37. Balance the equation to make ammonia:
    $$N_2 + H_2 \longrightarrow NH_3$$

38. The reaction between element A (black spheres) and element B (white spheres) is shown in the following diagram. Write the balanced chemical equation that best describes this reaction.

39. Hydrogen gas and carbon monoxide gas can be made to react to form methanol, CH_3OH. The diagram below represents a sample of carbon monoxide. Add hydrogen molecules to the drawing to represent the minimum amount of hydrogen needed to completely react with all of the carbon monoxide molecules.

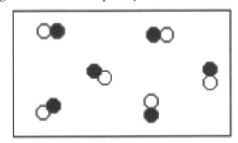

40. The active ingredient in marijuana is tetrahydrocannabinol (THC). The molecular formula for THC is $C_{21}H_{30}O_2$. a) Write the empirical formula for THC. b) What is the % by weight of carbon in THC?

41. Methanol, also known as wood alcohol, has the formula CH_3OH. a) What is the % by weight of carbon in methanol? b) How many grams of carbon are there in 16.0 g of methanol?

42. Write the complete and balanced chemical equation for the reaction of aqueous sodium chromate with aqueous silver nitrate.

43. Balance the following equations.
    a) $Fe_{(s)} + Cl_{2(g)} \longrightarrow FeCl_{3(s)}$
    b) $SiO_{2(s)} + C_{(s)} \longrightarrow Si_{(s)} + CO_{(g)}$
    c) $Fe_{(s)} + H_2O_{(g)} \longrightarrow Fe_3O_{4(s)} + H_{2(g)}$

44. Give the name for each of the following binary nonmetal compounds:
    a) BF_3
    b) P_2S_3
    c) OF_2

45. Complete and balance the following reactions:
    a) $C_4H_{10} + O_2 \longrightarrow$
    b) $Sr + O_2 \longrightarrow$
    c) $Mg + F_2 \longrightarrow$
    d) $PbCO_3 + heat \longrightarrow$

46. Complete and balance each of the following:
    a) $Ca_{(s)} + O_{2(g)} \longrightarrow$
    b) $Al_{(s)} + Br_{2(\ell)} \longrightarrow$
    c) $CH_3COOH_{(\ell)} + O_2 \longrightarrow$
    d) $C_6H_{14(\ell)} + O_2 \longrightarrow$

47. Give the formula for each of the following nonmetal compounds:
    a) bromine trifluoride
    b) xenon difluoride
    c) diphosphorus tetrafluoride

48. Write the formulas for each of the following:
    a) dinitrogen oxide
    b) sodium phosphate
    c) calcium chloride
    d) aluminum sulfate

49. Write the formula and formula weight (to the nearest amu) of each of the following:
    a) magnesium choride
    b) potassium bromate
    c) sodium phosphate

50. A 22.50 g mixture of sand and sodium chloride was mixed with water, heated to 90 °C and filtered. The student recovered 17.22 g of solid precipitate. What was the percent by weight of sodium chloride in the sample?

51. Automotive air bags inflate when solid sodium azide, NaN_3, rapidly decomposes to its component elements.
    a) Which one of the following diagrams correctly represents this reaction at room temperature?

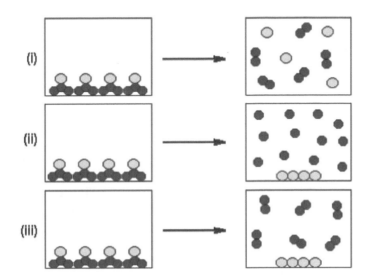

(I)

(ii)

(iii)

    b) How many moles of $N_{2(g)}$ are produced by the decomposition of 3.00 moles of $NaN_{3(s)}$?
    c) How many grams of $N_{2(g)}$ are produced by the decomposition of 2.45 g of $NaN_{3(s)}$?

52. Methane gas (CH_4) and chlorine gas react to form chloroform, $CHCl_{3(l)}$, and hydrogen chloride gas. The balanced chemical equation for the reaction is shown below. Consider the mixture of methane and chlorine shown in the diagram. Draw a representation of the product mixture assuming the reaction goes to completion.

$$CH_{4\,(g)} + 3\,Cl_{2\,(g)} \longrightarrow CHCl_{3\,(l)} + 3\,HCl_{(g)}$$

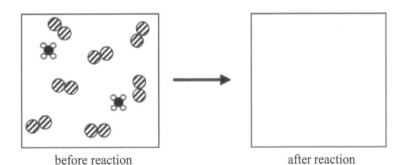

before reaction   after reaction

53. Hydrogen sulfide is composed of two elements: hydrogen and sulfur. In an experiment, 6.500 g of hydrogen sulfide are fully decomposed into hydrogen and sulfur.
    a) If 0.384 g of hydrogen is obtained in this experiment, how many grams of sulfur must be obtained?
    b) What fundamental law does this experiment illustrate?
    c) How is this law explained by Dalton's theory?

54. Nitrogen gas and hydrogen gas react to form ammonia, $NH_{3(g)}$. Consider the mixture of nitrogen and hydrogen shown in the diagram below. a) Draw a representation of the product mixture assuming the reaction goes to completion. b) What is the limiting reactant in this case? (c) How many grams of ammonia could be produced by the complete reaction of 3.50 g of nitrogen with 3.50 g of hydrogen?

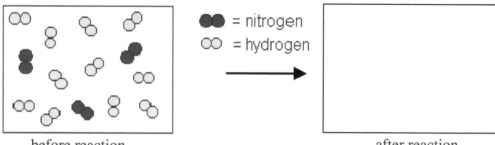

before reaction   after reaction

55. The diagram shown below on the left is a representation of solid sodium sulfate. a) In the box on the right, draw a representation of sodium sulfate dissolved in water (you do not need to draw in the water molecules). b) Explain why solid sodium sulfate does not conduct electricity but an aqueous solution of sodium sulfate does conduct electricity.

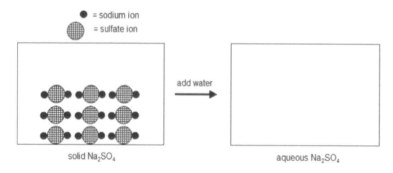

56. Name each of the following compounds, and indicate whether each is molecular or ionic:
   a) $KClO_3$
   b) SF_4
   c) $Co_2(SO_4)_3$
   d) P_2O_5

57. Circle the substances from the following list that are strong electrolytes.
   NaBr NO_2
   CH_4 CO_2
   $AgNO_3$ sulfur tetrafluoride
   calcium hydroxide H_2SO_4
   $FeCl_3$ cobalt(III) nitrate

58. You have a 0.12 M solution of $BaCl_2$. What ions exist in solution and what are their concentrations?

59. If some of a 0.100 M NaBr solution is spilled, what is the concentration of the solution left in the container?

60. What mass in grams of Na_2CO_3 is required for complete reaction with 25.0 mL of 0.155 M HNO_3?

$$Na_2CO_{3(s)} + 2\,HNO_{3(aq)} \longrightarrow 2\,NaNO_{3(aq)} + CO_{2(g)} + H_2O_{(\ell)}$$

61. There are several iron chlorides, most of which are hydrates. The formulas for four of the compounds are: $FeCl_2 - 2H_2O$, $FeCl_2 - 4H_2O$, $FeCl_3$, $FeCl_3 - 6H_2O$.
   a) Write the names of these compounds.
   b) A student made 150 grams of an iron chloride in the laboratory and found that it contained 42.1 g of Fe, 53.6 g of Cl, and 54.3 g of water. Which of the four iron chloride compounds did the student make?

62. Gaseous silane, SiH_4, ignites spontaneously in air according to the equation

$$SiH_{4(g)} + 2\,O_{2(g)} \longrightarrow SiO_{2(s)} + 2H_2O_{(g)}$$

If 5.2 L of $SiH_{4(g)}$ is treated with O_2, what volume of $O_{2(g)}$ is required for the complete reaction? What volume of $H_2O_{(g)}$ is produced? (Assume all gases are at the same temperature and pressure.)

63. A student reacts 30.0 g of benzene, C_6H_6, with 50.0 g of chlorine, Cl_2, to prepare chlorobenzene, C_6H_5Cl. The actual yield of chlorobenzene is 40.5 g. What is the percent yield for this reaction?

$$C_6H_6 + Cl_2 \longrightarrow C_6H_5Cl + HCl$$

64. A student combines 200 mL of a solution of 0.1 M magnesium nitrate with 100 mL of a solution of 0.05 M lithium hydroxide and obtains a white precipitate.
    a) What is the formula of the precipitate?
    b) Write the balanced molecular equation
    c) Write the balanced (complete) ionic equation
    d) Write the balanced net ionic equation
    e) What is the theoretical yield expressed in grams of the white precipitate?
    f) At the end of the reaction, how many moles of nitrate ion are present?

65. An aqueous solution of an unknown solute turns red litmus paper blue. The solution is only weakly conducting compared to a solution of NaCl of the same concentration. Which one of the following substances could be the unknown solute?

$$KOH, \ NH_3, \ HNO_3, \ KCl, \ H_3PO_4, \ CH_3CH_2CH_3$$

66. When several drops of aqueous NaOH are added to an unknown yellow aqueous solution, a reddish-brown precipitate forms. a) Which one of the following substances could be the unknown solute? b) Write the net ionic equation for the reaction that produced the precipitate.

$$AgNO_3, \ K_2CrO_4, \ NiCl_2, \ Na_2SO_4, \ FeCl_3$$

67. Write complete and balanced equations for the exchange reactions described below:
    a) aqueous sodium carbonate reacts with aqueous calcium nitrate
    b) aqueous copper(II) bromide reacts with aqueous sodium hydroxide

68. Balance the following combustion reaction of ethylene, C_2H_4, with oxygen:

$$C_2H_4 + O_2 \longrightarrow$$

    a) How many grams of O_2 are needed to burn 7.00 g of C_2H_4 according to the above equation?
    b) How many grams of carbon dioxide can form when a mixture of 5.25 g of ethylene and 3.75 g of oxygen react?
    c) How many grams of H_2O can be formed if 7.00 g of C_2H_4 and 20.00g of O_2 undergo complete combustion?

69. Octane, C_8H_{18}, a component of gasoline, undergoes a combustion reaction with oxygen.
    a) Complete and balance the equation for this reaction:

$$C_8H_{18} + O_2 \longrightarrow$$

    b) How many moles of O_2 are needed to burn 2.40 mol of C_8H_{18}?
    c) How many grams of O_2 are needed to burn 1.75 g of C_8H_{18}?
    d) Octane has a density of 0.692 g/mL at 20°C. How many grams of O_2 are required to burn 1.00 gal. of C_8H_{18}?
    e) Using the same density of octane given above, how many grams of CO_2 are produced by the combustion of 1.00 gal of C_8H_{18}?

70. What mass of NaOH is required to precipitate all the copper(II) ions from 100.0 mL of 0.100 M copper(II) nitrate solution?

71. Calculate the grams of H_2 gas in a 10 mL container at 36 °C and 550 mmHg pressure.

72. 23 mL of a gas at 45 °C and 720 mmHg is compressed to 15 mL and the temperature is raised to 50 °C. What is the final pressure of gas?

73. A mixture of gases at 760 torr contains 65.0% N_2, 15.0% O_2, and 20.0% carbon dioxide by volume. What is the partial pressure of each gas in torr?

74. A mass of neon gas occupies 200 cm³ at 100 °C. Find its volume at 0 °C, the pressure remaining constant. (Note: 1000 cm³ = 1 L)

75. One L of gas measured at 1 atm and −20 °C is compressed to 0.5 L when the temperature is 40 °C. What is the resulting pressure in units of atm after compression?

76. How high a column of air would be necessary to cause the barometer to read 76 cm of mercury, if the atmosphere were of uniform density 1.2 kg/m³? The density of mercury is 13.6×10^3 kg/m³.

77. A mass of oxygen occupies 5.00 L under a pressure of 740 torr. Determine the volume of the same mass of gas at standard pressure, the temperature remaining constant. (Standard pressure = 760 torr)

78. A mass of neon occupies 200 cm³ at 100 °C. Find its volume at 0 °C, the pressure remaining constant.

79. The atmosphere is a mixture of gases with a total pressure equal to the barometric pressure. A sample of the atmosphere at a total pressure of 740 mm Hg is analyzed to give the following partial pressures: $P(N_2)$ = 575 mm Hg, $P(Ar)$ = 6.9 mm Hg, $P(CO_2)$ = 0.2 mm Hg, $P(H_2O)$ = 4.0 mm Hg.
a) What is the partial pressure of $O_{2(g)}$ in this sample of the atmosphere?
b) What is the mole fraction of each gas in the sample?

80. At low temperatures and very low pressures, gases behave ideally, but as the pressure is increased the product of PV becomes less than the product of nRT. Give a molecular level explanation of this fact.

81. A gas phase reaction takes place in a syringe at a constant temperature and pressure. If the initial volume is 40 mL and the final volume is 60 mL, which of the following general reactions took place? Explain your reasoning.
a) $A_{(g)} + B_{(g)} \longrightarrow AB_{(g)}$
b) $2 A_{(g)} + B_{(g)} \longrightarrow A_2B_{(g)}$
c) $2 AB_{2(g)} \longrightarrow A_{2(g)} + 2 B_{2(g)}$
d) $2 AB_{(g)} \longrightarrow A_{2(g)} + B_{2(g)}$
e) $2 A_{2(g)} + 4 B_{(g)} \longrightarrow 4 AB_{(g)}$

82. Assuming that all of the containers shown below are at the same temperature, and that the volume of the container is proportional to the area of the 2D picture, which one is at the <u>highest</u> pressure? Explain your answer.

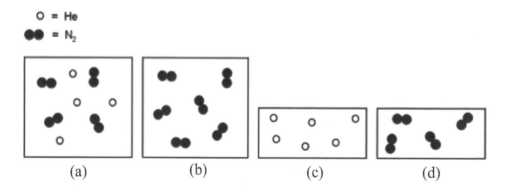

83. A scuba diver's tank contains 0.27 kg of O_2 compressed into a volume of 2.5 L.
a) Calculate the gas pressure inside the tank at 9 °C. (R = 0.0821 L·atm/K·mol)
b) What volume would this oxygen occupy at 24 °C and 750 torr?

84. Below is the equation for the combustion of methane (CH_4). How many liters of $H_2O_{(g)}$ are formed by the complete combustion of 1.8 grams of methane at 25 °C and 1.0 atm pressure?

$$CH_{4(g)} + 2\,O_{2(g)} \longrightarrow CO_{2(g)} + 2\,H_2O_{(g)}$$

85. The following data were collected by Robert Boyle in the late 1600's during his investigations of the relationship between pressure and volume for gases. (The data have been converted to modern metric units.) From his investigations, Boyle determined that the relationship between pressure and volume could be expressed by an equation:

$$P = \frac{k}{V}$$

Volume (L)	1/V	Pressure (atm)
0.0305	32.8	0.97
0.0292	34.2	1.02
0.0279	35.8	1.07
0.0267	37.5	1.12
0.0254	39.4	1.18
0.0241	41.4	1.24
0.0229	43.7	1.31
0.0216	46.3	1.39
0.0203	49.2	1.48
0.0191	52.5	1.57

A plot of P vs. $1/V$ produces a straight line. The slope of this line is the constant, k, in Boyle's equation. Today, we know that k is equal to nRT. a) Plot Boyle's data on the graph below. b) Assuming that Boyle did this experiment at room temperature, 25 °C, calculate the number of moles of gas used by Boyle in his experiment.

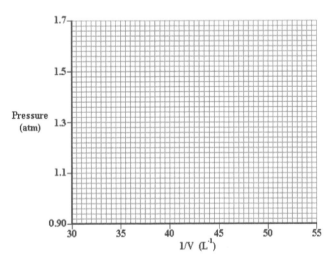

86. Fish need at least 4 ppm (= 4 mg L^{-1}) dissolved O_2 for survival in water.
   a) What is this concentration in mol L^{-1}?
   b) What partial pressure of O_2 above the water is needed to obtain this concentration at 10 °C? The Henry's Law constant at this temperature is 1.71×10^{-3} mol L^{-1} atm^{-1}. Remember Henry's Law is $C_g = k\,P_g$.

87. Place the following gaseous molecules in order of increasing average molecular speed at 300 K: CO_2, NO_2, HCl, O_2, F_2.

88. Assume the following gases behave ideally. a) Arrange them in order of increasing density at 1.00 atm and 298 K. b) Explain your answer.

    a) CO_2 b) N_2O c) Cl_2 d) O_2

    least dense ⟶ most dense

89. A mixture of 14.0 g of hydrogen, 84.0 g of nitrogen, and 2.0 moles of oxygen are placed in a flask at 25 °C. The partial pressure of the oxygen is 245 mm of mercury. What is the total pressure (in atm) in the flask?

90. The partial pressure of $O_{2(g)}$ in air is 21%. Calculate the molar concentration of oxygen in the surface water of a mountain lake at 20 °C and a total atmospheric pressure of 665 mmHg given the Henry's Law constant for oxygen is 1.38×10^{-3} mol/L-atm at this temperature.

91. A gas of unknown molecular mass was allowed to effuse through a small opening under constant pressure conditions. It required 85 s for 1.0 L of the gas to effuse. Under identical experimental conditions it required 22 s for 1.0 L of N_2 gas to effuse. Calculate the molar mass of the unknown gas.

92. You are given two flasks of equal volume. Flask A contains H_2 gas at 0 °C and 1 atm pressure. Flask B contains CO_2 gas at 0 °C and 2 atm pressure. Compare these two samples with respect to each of the following?
    a) average kinetic energy per molecule
    b) average molecular velocity
    c) number of molecules

93. Consider a system that releases heat to the surroundings and does work on the surroundings. a) Sketch a box to represent the system and draw labeled arrows to represent the heat and work transferred. b) Is it possible for ΔE to be positive for this process? Explain.

94. In principle, copper metal could be used to generate hydrogen gas (a valuable fuel) from water according the following equation.

$$Cu_{(s)} + H_2O_{(g)} \longrightarrow CuO_{(s)} + H_{2(g)}$$

    a) Is the reaction exothermic or endothermic?
    b) What is the enthalpy change (in kJ) for the reaction if 2.00 g of copper metal reacts with excess water vapor at constant pressure?

95. Calculate ΔE for the following processes and determine whether they are exothermic or endothermic:
    a) A system absorbs 113 kJ of heat from the surroundings and does 39 kJ of work.
    b) The system absorbs 77.5 kJ of heat while the surroundings do 63.5 kJ of work on the system.

96. Calculate the overall change in thermal energy for the following solution process:
    ΔH separation of solute = 23J
    ΔH separation of solvent = 45J
    ΔH formation of solvent solute intermolecular forces = −71J

97. The incomplete combustion of acetic acid, $CH_3CO_2H_{(l)}$, to form liquid water and gaseous carbon monoxide at constant pressure releases 305.7 kJ of heat per mole of acetic acid to the surroundings.
    a) Write a balanced thermochemical equation for this reaction.
    b) Is the reaction exothermic or endothermic?

98. The specific heat of copper metal is 0.385 J/g·°C. How many joules of energy are required to lower the temperature of 3.67 g of copper from 76.2 °C to 23.0 °C?

99. When a 5.60 g sample of solid sodium hydroxide dissolves in 100.0 g of water in a coffee-cup calorimeter, the temperature rises from 22.3 °C to 41.6 °C. Calculate $\underline{\Delta H}$ $\underline{\text{in kJ/mol}}$ for the solution process if the specific heat capacity of the solution is 4.18 J/g·K.

$$NaOH_{(s)} \longrightarrow Na^+_{(aq)} + OH^-_{(aq)}$$

100. A sample of 100 g of gold, originally at 100 °C, is plunged into 100 g of water originally at 25 °C. The final temperature of the mixture is 27.2 °C. In this experiment, the metal has the _____ heat capacity because it underwent the _____ temperature change.
    a) larger, larger   b) larger, smaller   c) smaller, larger   d) smaller, smaller

101. Consider the graph shown below. Which substance has the highest specific heat capacity? Why?

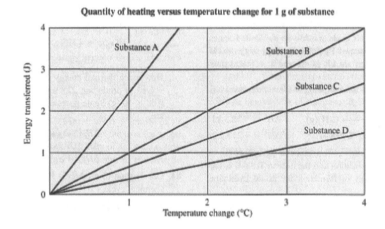

Quantity of heating versus temperature change for 1 g of substance

102. One way to cool a cup of coffee is to plunge an ice-cold piece of aluminum into it. Suppose a 20.0 g piece of aluminum is stored in the refrigerator at 0.0 °C and then dropped into a cup of coffee. The coffee's temperature drops from 90.0 °C to 75.0 °C. How many kJ of thermal energy did the piece of aluminum absorb? (C_{Al} = 0.902 J/g·°C, C_{water} = 4.18 J/g·°C).

103. When a 12.50 g sample of solid calcium hydroxide dissolves in 100.0 g of water in a coffee-cup calorimeter, the temperature rises from 24.1°C to 45.5 °C. Calculate ΔH in kJ/mol for the solution process if the specific heat of the solution is 4.18 J/g·K.

$$Ca(OH)_{2(s)} \longrightarrow Ca^{2+}_{(aq)} + 2\,OH^-_{(aq)}$$

104. Aluminum and iron have specific heats of 0.90 J/g·K and 0.45 J/g·K respectively. If identical amounts of heat were added to an aluminum bar and an iron bar which have identical masses, the temperature of which bar would increase less?

105. The sketch below shows two identical beakers with different volumes of water at the <u>same temperature</u>.
    a) Is the thermal energy content of beaker 1 greater than, less than, or equal to that of beaker 2? Explain your reasoning.

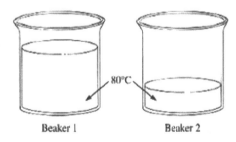

Beaker 1   Beaker 2

    b) If 150 J of thermal energy were transferred to each beaker, would the temperature of beaker 1 be greater than, less than, or equal to that of beaker 2? Explain your reasoning.

106. From the following enthalpies of reaction:

$$H_{2(g)} + F_{2(g)} \longrightarrow 2\ HF_{(g)} \qquad\qquad \Delta H = -537\ kJ$$

$$C_{(s)} + 2\ F_{2(g)} \longrightarrow CF_{4(g)} \qquad\qquad \Delta H = -680\ kJ$$

$$2\ C_{(s)} + 2\ H_{2(g)} \longrightarrow C_2H_{4(g)} \qquad\qquad \Delta H = +52.3\ kJ$$

calculate ΔH for the reaction of ethylene with F_2:

$$C_2H_{4(g)} + 6\ F_{2(g)} \longrightarrow 2\ CF_{4(g)} + 4\ HF_{(g)}$$

107. When 10.0 g of nitrogen dioxide are formed from the combustion of nitrogen, 7.36 kJ of heat are absorbed from the surroundings.
   a) Is this reaction is endothermic or exothermic?
   b) Calculate the enthalpy change for the reaction $N_{2(g)} + 2\ O_{2(g)} \longrightarrow 2\ NO_{2(g)}$

108. Circle the letter of each reaction below for which the enthalpy change of the reaction is also the enthalpy of formation of the product.
   a) $C_{(s)} + 2\ F_{2(g)} \longrightarrow CF_{4(g)}$
   b) $H_{2(g)} + F_{2(g)} \longrightarrow 2\ HF_{(g)}$
   c) $SF_{4(g)} + F_{2(g)} \longrightarrow SF_{6(g)}$

109. Using the values from the table below, calculate the standard enthalpy change for each of the following reactions:

Substance	ΔH_f° (kJ/mol)
$SO_{2(g)}$	-270
$SO_{3(g)}$	-395
$Mg(OH)_{2(s)}$	-925
$MgO_{(s)}$	-602
$H_2O_{(g)}$	-242
$H_2O_{(\ell)}$	-286

   a) $2SO_{2(g)} + O_{2(g)} \longrightarrow 2\ SO_{3(g)}$
   b) $Mg(OH)_{2(s)} \longrightarrow MgO_{(s)} + H_2O_{(\ell)}$

110. Calculate the enthalpy change when 50.0 g of graphite are burned according to the following equation. Is this reaction endothermic or exothermic?

$$2\ C\ (graphite) + O_2 \longrightarrow 2\ CO_{(g)} \qquad\qquad \Delta H^\circ = -221.0\ kJ$$

111. Glucose is a common sugar with the formula $C_6H_{12}O_6$. Determine the molar enthalpy of combustion for glucose (in kJ/mol).

$$C_6H_{12}O_{6(s)} + 6\ O_{2(g)} \longrightarrow 6\ CO_{2(g)} + 6\ H_2O_{(g)}$$

Substance	ΔH_f° (kJ/mol)
$C_6H_{12}O_{6(s)}$	-1273
$CO_{2(g)}$	-394
$H_2O_{(g)}$	-242

112. Using the following reactions, find the standard enthalpy change for the formation of 1.00 mol of $PbO_{(s)}$ from lead metal and oxygen gas.

$$PbO_{(s)} + C \text{ (graphite)} \longrightarrow Pb_{(s)} + CO_{(g)} \qquad\qquad \Delta H° = +106.8 \text{ kJ}$$

$$2 \text{ C (graphite)} + O_2 \longrightarrow 2 CO_{(g)} \qquad\qquad \Delta H° = -221.0 \text{ kJ}$$

113. Write the equation that represents the heat of combustion of methanol, $CH_3OH_{(\ell)}$. Given the following information, determine the heat of combustion of methanol.

$$C_{(s)} + 2 H_{2(g)} + \tfrac{1}{2} O_{2(g)} \longrightarrow CH_3OH_{(\ell)} \qquad\qquad \Delta H = -239.2 \text{ kJ}$$

$$H_{2(g)} + \tfrac{1}{2} O_{2(g)} \longrightarrow H_2O_{(\ell)} \qquad\qquad \Delta H = -285.8 \text{ kJ}$$

$$C_{(s)} + O_{2(g)} \longrightarrow CO_{2(g)} \qquad\qquad \Delta H = -393.5 \text{ kJ}$$

114. The enthalpies of formation for $O_{2(g)}$ and $H_2O_{(\ell)}$ are 0.0 kJ/mole and -286 kJ/mol respectively. What is the enthalpy of formation (in kJ/mol) for $H_2O_{2(\ell)}$?

$$2 H_2O_{2(\ell)} \longrightarrow O_{2(g)} + 2 H_2O_{(\ell)} \qquad\qquad \Delta H = -196 \text{ kJ}$$

115. For each of the following pairs, indicate which substance possesses the larger standard entropy.
    a) 50 g of $NaCl_{(s)}$ at 25 °C or 50 g of $NaCl_{(aq)}$ at 50 °C
    b) 1 mol of $C_3H_{8(g)}$ at 25 °C or 1 mol of $C_2H_{6(g)}$ at 25 °C
    C) 1 mol of $N_{2(g)}$ at 25 °C in a 1 L container or 1 mol of $N_{2(g)}$ at 25 °C in a 4 L container

116. Predict the sign of the entropy change for the following reactions:

$$C_2H_{4(g)} + H_{2(g)} \longrightarrow C_2H_{6(g)}$$

$$4 Al_{(s)} + 3 O_{2(g)} \longrightarrow 2 Al_2O_{3(s)}$$

117. Calculate the heat transfer when 5.0 g of ice at -5.0 °C is heated to 120.0 °C. (Use specific heats and ΔH's from class notes)

118. Calculate the kinetic energy in Joules of a 156 g (5.50 oz.) baseball moving at 44.7 m/s (100 mph).

119. A balloon is heated by adding 900 kJ of heat. It expands, doing 400 kJ of work on the surrounding atmosphere.
    a) Calculate the change in internal energy of the system (balloon and its contents).
    b) Is this process endothermic or exothermic?

120. Predict the sign of $\Delta S°$ for each of the following processes.
    a) $2 SO_{3(g)} \longrightarrow 2 SO_{2(g)} + O_{2(g)}$
    b) $Pb^{2+}_{(aq)} + SO_4^{2-}_{(aq)} \longrightarrow PbSO_{4(s)}$
    c) $Ca(OH)_{2(s)} \longrightarrow CaO_{(s)} + H_2O_{(g)}$

121. Hydrazine (N_2H_2) and 1,1-dimethylhydrazine ($N_2H_2(CH_3)_2$) both react spontaneously with O_2 and can be used as rocket fuels. The value of $\Delta H_f°$ of liquid hydrazine is $+50.6$ kJ/mole, and that of liquied dimethylhydrazine is $+49.2$ kJ/mol. Decide whether the reaction of hydrazine or dimethylhydrazine with oxygen gives more heat per gram (at constant pressure). (Hint: you will have to determine the value of $\Delta H°$ for each of the reactions shown below.)

$$N_2H_{2(\ell)} + O_{2(g)} \longrightarrow N_{2(g)} + 2 H_2O_{(g)}$$

$$N_2H_2(CH_3)_{2(\ell)} + 4 O_{2(g)} \longrightarrow 2 CO_{2(g)} + 4 H_2O_{(g)} + N_{2(g)}$$

122. Which member of each of the following pairs has the lower entropy? Explain.
    a) 1 mol of each of $Fe_{(l)}$ or $Fe_{(s)}$
    b) 1 mol of each of $CH_{4(g)}$ or $C_2H_{6(g)}$

123. Explain why the entropy of the system increases when solid $NaCl_{(s)}$ dissolves in water.

124. Calculate the value of $\Delta G°$ for the following reaction at 25 °C. Indicate whether the reaction will be: a) spontaneous at all temperatures, b) non-spontaneous at all temperatures, c) spontaneous at low temperatures only, or d) spontaneous at high temperatures only.

   a) $N_{2(g)} + 3\ F_{2(g)} \longrightarrow 2\ NF_{3(g)}$ $\Delta H° = -249$ kJ, $\Delta S° = -278$ J/K

   b) $N_2F_{4(g)} \longrightarrow 2\ NF_{2(g)}$ $\Delta H° = 85$ kJ, $\Delta S° = 198$ J/K

125. Without using tables to calculate values, predict the signs for ΔH, ΔS, and ΔG for the following processes:

$$C_2H_{4(g)} + 3\ O_{2(g)} \longrightarrow 2\ CO_{2(g)} + 2\ H_2O_{(\ell)}$$

$$Fe_{(s)} \longrightarrow Fe_{(\ell)}\ \text{at } 25°C$$

126. For a particular reaction, $\Delta H = -45$ kJ and $\Delta S = -101$ J/K. Assume that ΔH and ΔS do not change very much with temperature. a) At what temperature will $\Delta G = 0$? b) If the temperature is increased from what was calculated in part (a), will the reaction be spontaneous or non-spontaneous?

127. Glucose is a common sugar with the formula $C_6H_{12}O_6$. Determine the standard free energy change ($\Delta G°$) for the combustion of glucose.

$$C_6H_{12}O_{6(s)} + 6\ O_{2(g)}\ 6\ CO_{2(g)} + 6\ H_2O_{(g)}$$

Substance	$\Delta G_f°$ (kJ/mol)
$C_6H_{12}O_{6(s)}$	−910
$CO_{2(g)}$	−394
$H_2O_{(g)}$	−229

128. Consider the following reaction:

$$H_2O_{(\ell)} \longrightarrow H_2O_{(g)}\ \Delta H° = 44\ \text{kJ};\ \Delta S° = 118.9\ \text{J/K}$$

   a) Calculate $\Delta G°$ at 25 °C.
   b) Is this reaction spontaneous at room temperature?
   c) Of the following conditions indicate (circle) which is appropriate:
      i. Reaction occurs
      ii. Reaction doesn't occur
      iii. Reaction only occurs at high temperatures

129. Pure oxygen can be produced by the decomposition of potassium chlorate to oxygen and potassium chloride by the following reaction:

$$2\ KClO_{3(s)} \longrightarrow 2\ KCl_{(s)} + 3\ O_{2(g)}$$

Substance	$\Delta H_f°$ (kJ/mol)	$\Delta G_f°$ (kJ/mol)	$S°$ (J/mol-K)
$KClO_{3(s)}$	−391.2	−289.9	143.0
$KCl_{(s)}$	−435.9	−408.3	82.7
$O_{2(g)}$	0	0	205.0

   a) What are the enthalpy, entropy, and free energy changes for the reaction shown above at 25 °C?
   b) Calculate the free energy changes using values for $\Delta G_f°$, and then again using the formula $\Delta G = \Delta H - T\Delta S$.
   c) Is the reaction spontaneous (product favored) or non spontaneous (reactant favored)? Explain.

130. The energy required to break the bond between two oxygen atoms in an oxygen molecule is 8.2×10^{-19} J. What wavelength of electromagnetic radiation possesses enough energy to break this bond?

131. a) What is the frequency of electromagnetic radiation that has a wavelength of 610 pm?
    b) The visible portion of the electromagnetic spectrum covers the range from approximately 350 to 700 nm. Is the electromagnetic radiation described in part a) visible to the eye?
    c) How far would the electromagnetic radiation described in part a travel in 1.0 second?

132. What is the energy in J of a dozen photons with a wavelength of 200 nm?

133. It requires 183.7 kJ/mol to eject electrons from cesium metal.
    a) What is the minimum frequency of light necessary to emit electrons from cesium via the photoelectric effect?
    b) What is the wavelength of this light
    c) Is this light visible to the human eye (between 350–700 nm)?

134. The diagram below represents a 1 nanosecond snapshot of electromagnetic radiation. What is the frequency (Hz) of this radiation?

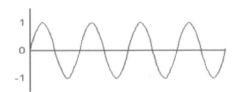

135. It requires a photon with a minimum energy of 4.41×10^{-19} J to emit electrons from sodium.
    a) What is the minimum frequency of light necessary to emit electrons from sodium via the photoelectric effect?
    b) What is the wavelength of this light?
    c) Is this light visible to the human eye (between 350–700 nm)?

136. Which of the following contains the most total energy?
    a) 4000 photons of red light ($\lambda = 650$ nm)
    b) A billion radiowave photons ($\nu = 100$ mHz)
    c) One X-ray photon ($E = 6.63 \times 10^{-16}$ J)
    d) A dozen uv photons ($\lambda = 5 \times 10^{-8}$ m)

137. Consider a 400 nm (blue) photon and a 650 nm (red) photon.
    a) Which photon has the higher frequency?
    b) Which photon has less energy?
    c) Which photon has a higher velocity in a vacuum?

138. Which sunscreen is the most effective at blocking the <u>highest energy</u> radiation?

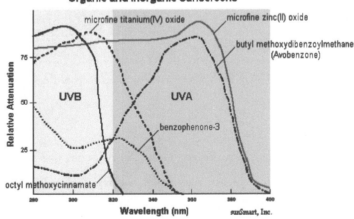

139. The US Navy has a system for communicating with submerged submarines. The system uses radio waves with a frequency of 76 Hz. What is the wavelength of this radiation in kilometers?

140. Assume a microwave oven operates at a frequency of 1.00×10^{11} Hz.
    a) What is the wavelength of this radiation in meters?
    b) What is the energy in joules per photon?
    c) What is the energy per mole of photons?

141. The Lyman series of emission lines from hydrogen atoms are those for which $n_f = 1$.
    a) Calculate the wavelength of the second line in the Lyman series ($n_i = 3$).
    b) What region of the electromagnetic spectrum is this?

$$\Delta E = -2.18x10^{-18} \text{ J} \left(\frac{1}{n_f^2} - \frac{1}{n_i^2}\right)$$

142. Is energy absorbed or emitted when an electron in a hydrogen atom moves from n=4 level to n=2 level? Explain. How many lines could be observed in the transition from n=4 orbit to n=2 orbit in a hydrogen atom? Explain.

143. What is the energy in J of a dozen photons with a wavelength of 200 nm?

144. Which of the following sets of quantum numbers or electron configurations cannot exist in a hydrogen atom? Why?
    a) $n = 2, l = 2, m_l = 0$ b) $n = 3, l = 0, m_l = -1$
    c) $1p^3$ d) $3f^6$

145. Sketch the shape and orientation of the following types of orbitals: a) s, b) p_x, c) d_{z^2}

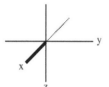  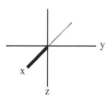

146. The following orbital diagram represents a violation of

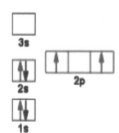

    (a) Pauli Exclusion Principle
    (b) Aufbau Principle
    (c) Hund's Rule
    (d) None of these

147. Draw orbital diagrams (boxes with arrows) representing the <u>valence electrons</u> for each of the following atoms or ions. Use the symbols of the nearest noble gases to represent the core electrons.
    a) Cl
    b) Zn^{2+}

148. Identify the group of elements (1A, 2A, etc.) that corresponds to each of the following generalized electron configurations:
    a) [noble gas]ns^2np^4
    b) [noble gas]$ns^2(n-1)d^3$

149. Which will be closer to the nucleus, the n=3 electron shell in Ar or the n=3 shell in Kr? Explain.

150. Notice that among the nonmetallic elements, the change in atomic radius in moving one place left or right in a row is smaller than the change in moving one row up or down. Explain these observations.

151. The calculated non-bonding radius of a hydrogen atom is 53 pm and for a carbon atom it is 67 pm. a) Estimate the length of a C-H covalent bond from these data. b) The average measured length of a C-H covalent bond is 110 pm. Explain why this is different than what you estimated in part (a).

152. a) Predict whether each of the following oxides is ionic or molecular. b) Predict which of the oxides will form acidic solutions with water and which will form basic solutions.
   i. SO_3
   ii. Na_2O_2
   iii. P_2O_5
   iv. MgO

153. The elements in group 6A of the periodic table are known as the calcogens (sometimes spelled "chalcogens"). Write a balanced chemical equation for the reaction of a calcogen with an alkaline earth metal.

154. If the four quantum numbers for an electron in a 2p orbital (in the ground state) are $n=2$, $l=1$, $m_\ell = -1$, and $m_s = +1/2$ and there are no other electrons in the 2p orbitals, what would be the 4 quantum numbers for the second electron to be put into a 2p orbital (ground state)?

155. Provide the name and the number of subshells for each of the following electron shells:
   a) $n=4$
   b) $n=3$

156. Which of the following cannot exist:
   1s, 2p, 1p, 3d, 2d, 3f

157. Write the Lewis symbols for the following atoms or ions. a) Mg, b) Ar, c) Al^{3+}, d) Br^-.

158. a) How many electrons must an oxygen atom gain to achieve an octet in its valence shell? b) If an atom has the electron configuration $1s^2 2s^2 2p^6 3s^2 3p^5$, how many electrons must it gain to achieve an octet?

159. Draw Lewis structures for each of the following molecules or ions:
   a) CH_2Cl_2
   b) OF_2
   c) SO_3^{2-}

160. Do any of the species listed in the problem above exhibit resonance? If so, draw the different resonance forms that could exist.

161. Draw the Lewis dot structure for the hypoiodite anion.

162. Write the Lewis symbols for the following atoms or ions. a) Mg b) Ar c) Al^{3+} d) Br^-.

163. Draw the Lewis dot structures for CH_2Cl_2, PCl_5, PO_4^{3-}, NH_4^+, CN^-

164. Draw the Lewis dot structures for CO_2, HCO_3^-, NF_3, ClO_3^-, IF_5,

165. For each of the following Lewis symbols, give an element that could correctly be put in place of X.
   a) •X•
   b) •X
   c) : X :

166. Which pair of bonding atoms would have the shortest bond length? Which would have the longest?
   C-C   C-O   C-P   C-F   C-Si

167. Which pair of bonding atoms would have the strongest bond? Which would have the weakest?
   F-Cl   N-Cl   Cl-Cl

168. Determine which of the following covalent bonds would be considered to be polar:
   a) Br—Br
   b) O—Cl
   c) S—F
   d) P—O

169. Circle the molecule in each pair that has the **most** polar bonds.
    a) SO_2 or SF_4
    b) OF_2 or BrO_2
    c) KCl or PCl_3

H 2.1																	
Li 1.0	Be 1.5											B 2.0	C 2.5	N 3.0	O 3.5	F 4.0
Na 0.9	Mg 1.2											Al 1.5	Si 1.8	P 2.1	S 2.5	Cl 3.0
K 0.8	Ca 1.0	Sc 1.3	Ti 1.5	V 1.6	Cr 1.6	Mn 1.5	Fe 1.8	Co 1.9	Ni 1.9	Cu 1.9	Zn 1.6	Ga 1.6	Ge 1.8	As 2.0	Se 2.4	Br 2.8
Rb 0.8	Sr 1.0	Y 1.2	Zr 1.4	Nb 1.6	Mo 1.8	Tc 1.9	Ru 2.2	Rh 2.2	Pd 2.2	Ag 1.9	Cd 1.7	In 1.7	Sn 1.8	Sb 1.9	Te 2.1	I 2.5
Cs 0.7	Ba 0.9	Lu 1.2	Hf 1.3	Ta 1.5	W 1.7	Re 1.9	Os 2.2	Ir 2.2	Pt 2.2	Au 2.4	Hg 1.9	Tl 1.8	Pb 1.9	Bi 1.9	Po 2.0	At 2.2
Fr 0.7	Ra 0.9	Ac 1.1	Th 1.3	Pa 1.4	U 1.4											

Electronegativity Values of the Elements

170. How does atomic size and electronegativity vary with respect to position on the periodic table? Why do they vary this way?

171. Compared to a carbon atom, an oxygen atom is. . .
    a) bigger, has higher electronegativity, and is more metallic
    b) smaller, has higher electronegativity, and is less metallic.
    c) smaller, has lower electronegativity, and more metallic
    d) bigger, has lower electronegativity, and is more metallic.

172. Arrange the following atoms in order of increasing atomic radius: Cl, Ge, Ca, Sr.

173. Arrange the following elements in order of increasing atomic size: Al, B, C, K, and Na.

174. Write the electronic configuration for N.

175. a) How many valence electrons does a carbon atom have? b) An atom has the electron configuration $1s^2 2s^2 2p^6 3s^2 3p^3$. c) How many valence electrons does it have? d) What element is described in part b)?

176. Write the complete ground state electron configuration for the following. Indicate the number of valence electrons for each atom.
    a) V
    b) Cu

177. Write the complete ground state electron configurations for the following species:
    a) Mg
    b) Mg^{2+}

178. Write the complete electron configurations for the following ions, indicating the number of valence electrons in each. Indicate which ions are isoelectronic.
    a) Na^+
    b) Al^{3+}
    c) Cl

179. Fill in the boxes to show the electronic configuration of Si and label each box to signify the orbital (ex. 1s or 3d):

□   □   □□□   □   □□□

180. Fill in the boxes to show the electronic configuration of N and label each box to signify the orbital (ex. 1s or 3d):

☐ ☐ ☐☐☐

181. Why is the second ionization energy of Na much greater than the second ionization energy of Mg?

182. Arrange the following atoms in order of increasing first ionization energy: F, Al, P, and Mg.

183. Rank the following ionization energies (IE) in order from smallest to the largest:
a) first IE of Be
b) first IE of Li
c) second IE of Be
d) second IE of Na
e) first IE of K

184. The first ionization energy of the halogens decreases as you go down the group in the periodic table. Explain how this trend relates to the variation in atomic radii.

185. Circle the atom with the smaller atomic radius in the following:
a) Na or Ar
b) Li or Na
c) Li or Be

186. Is energy absorbed or emitted when an electron in a hydrogen atom moves from n=4 level to n=2 level? Explain. How many lines could be observed in the transition from n=4 orbit to n=2 orbit in a hydrogen atom? Explain.

187. Draw Lewis dot structures for the following molecules or ions. Determine the electron pair and molecular geometry for each? Indicate if the molecule is polar or non-polar.
a) NH_2Cl
b) OCl_2
c) ClF_3
d) ClF_4^-
e) ClF_5

188. The transition in hydrogen that results in the greatest release of energy is
a) n= 1 to n= 3
b) n= 2 to n=6
c) n=5 to n=2
d) n=2 to n=1

189. Ca has atomic number 20. The 20th electron goes into what type of orbital: s, p, d, or f?

190. Mn has atomic number 25. The 25th electron goes into what type of orbital: s, p, d, or f?

191. Select the atom or ion in each of the following pairs that has the larger radius. Explain your selection.
a) Cs or Rb
b) O^{2-} or O
c) Br or As
d) S^{2-} or P^{3-}
e) K^+ or Rb^+
f) Cu^{2+} or Cu^+

192. While the electron affinity for fluorine is a negative quantity, it is positive for neon. Explain this difference in terms of the electron configurations of the two elements.

193. Predict the bond angles and polarity of each molecule. Some molecules will have more than one unique bond angle, so clearly indicate which ones you are describing. If the molecule is polar, draw an arrow on its structure indicating the direction of the dipole moment. In the last column, circle the type of hybrid orbitals used for bonding by the central atom. The first row of the table is completed for you.

Name	Molecular Formula	Ball and Stick Renderings	Bond Angles	Polarity	Hybrid Orbitals used for Bonding by Central Atom
Oxygen Difluoride	OF_2		F-O-F 105°	[X] polar [ ] nonpolar	sp sp² (sp³) sp³d sp³d²
Boron Trifluoride	BF_3			[ ] polar [ ] nonpolar	sp sp² sp³ sp³d sp³d²
Trifluoroiodo -methane	CF_3I			[ ] polar [ ] nonpolar	sp sp² sp³ sp³d sp³d²
Phosphorus Pentafluoride	PF_5			[ ] polar [ ] nonpolar	sp sp² sp³ sp³d sp³d²
Sulfur Tetrafluoride	SF_4			[ ] polar [ ] nonpolar	sp sp² sp³ sp³d sp³d²
Ammonia	NH_3			[ ] polar [ ] nonpolar	sp sp² sp³ sp³d sp³d²

Name	Molecular Formula	Balls and Stick Renderings	Bond Angles	Polarity	Hybrid Orbitals used for Bonding by Central Atom
Xenon Tetrafluoride	XeF_4			☐ polar ☐ nonpolar	sp sp^2 sp^3 sp^{3}d sp^{3}d^2
Carbon Dioxide	CO_2			☐ polar ☐ nonpolar	sp sp^2 sp^3 sp^{3}d sp^{3}d^2
Formaldehyde	CH_2O			☐ polar ☐ nonpolar	sp sp^2 sp^3 sp^{3}d sp^{3}d^2
Acetic Acid	CH_3CO_2H		O-C-O	☐ polar ☐ nonpolar	sp sp^2 sp^3 sp^{3}d sp^{3}d^2

Indicate hybrid orbitals used for bonding by this carbon atom

194. Consider the following substances: CH_2Cl_2, CCl_4, H_2O. Which has a) the largest London dispersion forces; b) no dipole-dipole forces.

195. Describe the intermolecular forces that must be overcome to convert each of the following from a liquid to a gas: a) CO_2 b) CH_3F c) NCl_3 d) PH_3 e) CH_3OH

196. Of the following substances, which has London dispersion forces as its **only** intermolecular force? Why?
a) CH_3OH b) NH_3 c) H_2S d) CH_4

197. Which of the following substances use London dispersion as one of their intermolecular forces. Explain.
a) CH_3OH b) NH_3 c) H_2S d) CH_4 e) all of the above

198. Indicate whether the following substances will be more soluble in H_2O or C_8H_{18}:
a) K_3PO_4
b) CH_4
c) CH_3OH
d) C_2H_2

199. Circle any polar compounds below
a) CH_2F_2
b) Cl_2
c) SiF_4
d) BI_3
e) PH_3

200. What intermolecular forces exist between the following pairs of molecules? List all that apply.
   a) CO and NH_3
   b) CO_2 and Cl_2
   c) K^+ and H_2O
   d) NH_3 and CH_4
   e) CH_2O and H_2O

201. Determine whether the molecules below are polar or non-polar. You will need to draw Lewis dot structure and determine the molecular geometry and electron domain geometry first.
   a) Cl_2
   b) HCN
   c) CF_4
   d) CF_2Cl_2
   e) $BClF_2$

202. What intermolecular forces exist between each pair below?
   a) Br_2 and F_2
   b) CH_4 and H_2O
   c) Na^+ and H_2O
   d) NH_3 and CF_4

203. Which substance in each pair is more soluble in water? Explain.
   a) $CH_3CH_2CH_2CH_2Cl$ or $CH_3CH_2CH_2CH_2OH$
   b) $CaBr_2$ or CBr_4
   c) C_6H_6 (benzene) or C_6H_5OH (phenol)

204. Which of the following would you expect to be most soluble in cyclohexane (C_6H_{12})? The least soluble? Explain your reasoning.
   a) NaCl
   b) CH_3CH_2OH
   c) C_3H_8

205. Use the following information about ethanol to calculate the amount of heat (in kJ) that would be required to raise the temperature of 50.0 g of ethanol from −125 °C to 35.0 °C.

   melting point = −114 °C   boiling point = 78 °C
   ΔH_{fus} = 5.0 kJ/mol ΔH_{vap} = 38.6 kJ/mol
   c_{solid} = 0.97 J/g·K c_{liquid} = 2.3 J/g·K

206. Based on their critical temperatures, which of the following substances can be liquefied at room temperature (25 °C)? Circle your choices.

Substance	Critical Temperature (K)
Ammonia, NH_3	405.6
Argon, Ar	150.9
Water, H_2O	647.6
Oxygen, O_2	154.4
Carbon Dioxide, CO_2	304.3

207. ***Without looking at your notes***, construct a phase diagram for a substance. Label each axis. Indicate where the solid liquid and gas phases are located. With arrows, show and label all 6 phase changes described in class. Indicate where the normal boiling point and normal melting point is located (be sure this corresponds to actual pressures and temperatures on each axis. Identify the triple point.

208. Use the phase diagram shown below to describe all of the phase changes that would occur in each of the following cases for this substance. a) The substance, originally at 0.1 atm and −10 °C, is slowly compressed at constant temperature until the final pressure is 5 atm. b) The substance, originally at 50 °C and 5 atm, is cooled at constant pressure until the temperature is −10 °C.

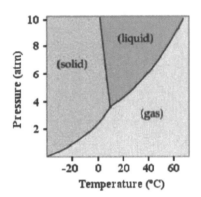

209. Liquid ammonia, $NH_{3(l)}$, was once used in home refrigerators as the heat transfer fluid. The specific heat of the liquid is 4.7 J/g·°C and that of the vapor is 2.2 J/g·°C. The enthalpy of vaporization is 23.33 kJ/mol at the boiling point.
    a) If you heat 12 kg of liquid ammonia from −50.0 °C to its boiling point of −33.3 °C, allow it to evaporate, and then continue warming on to 0.0 °C, how much heat energy must you supply?
    b) Sketch the heating curve for this process.

210. The molar heat of vaporization of methanol is 38.0 kJ/mol at 25 °C. How much heat is required to convert 250 mL of methanol form liquid to vapor? The density of methanol is 0.787 g/mL at 25 °C.

211. The energy required to expand the surface area of a liquid by a unit amount of area is known as the liquid's
    a) viscosity   b) surface tension   c) volatility   d) meniscus

212. Which of the following liquids has the highest surface tension? Explain.
    a) H_2O b) $CH_3CH_2-O-CH_2CH_3$ c) $H_2C=O$

213. Circle all of the factors that will affect the vapor pressure of a liquid.
    a) volume of the liquid
    b) surface area of the liquid
    c) intermolecular attractive forces
    d) temperature

214. PCl_3 is more volatile than $AsCl_3$ at 25 °C.
    a) Which substance has the greater intermolecular attractive forces?
    b) Which substance has the higher vapor pressure?
    c) Which substance will have the higher boiling point?
    d) Which substance is most likely to have the highest viscosity?

215. Which of these substances has the highest melting point? The lowest melting point? Explain your answers.
    a) LiBr
    b) CaO
    c) CO
    d) CH_3OH

216. The heat of fusion of water is 6.01 kJ/mol. The heat capacity of liquid water is 75.2 J/mol·K. How much heat is required to convert 50.0 g of ice at 0 °C to liquid water at 22 °C?

217. The triple point and critical points in the phase diagram below are located at (respectively):
a) 2 and 4   b) 2 and 3   c) 1 and 2   d) 3 and 4

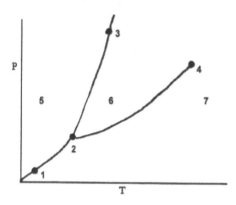

218. Aluminum metal crystallizes with a face-centered cubic unit cell. a) How many Al atoms are there in a unit cell of the crystal? b) Assume that the aluminum atoms can be represented as spheres as shown below. If each Al atom has a radius of 143 pm, what it the length of a side of the unit cell?

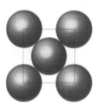

219. A metal crystallizes in a simple cubit unit cell structure. The length of the edge of the cubic unit cell is 420 pm. What is the radius of the meal atoms?

220. A metal crystallizes in a body centered cubic unit cell structure. The radius of the atom is 300 pm. What is the length of the edge of the cubic unit cell?

221. For each of the following pairs of substances, predict which substance would have the higher melting point. Explain our predictions.
a) Ar, Xe
b) SiO_2, CO_2
c) KBr, Br_2
d) CCl_4, C_6H_6

222. Calculate the overall change in thermal energy for the following solution process:
ΔH separation of solute = 23 J
ΔH separation of solvent = 45 J
ΔH formation of solvent solute intermolecular forces = −71 J

223. For an endothermic solution process, will decreasing the temperature increase or decrease the solubility of the solute in the solvent?

224. For an exothermic solution process, will decreasing the temperature increase or decrease the solubility of the solute in the solvent?

225. Arrange the following aqueous solutions in order of <u>decreasing</u> vapor pressure of water at 25 °C.
a) 0.15 M KBr   b) 0.15 M $CaCl_2$ c) 0.25 M sucrose   d) 0.10 M $Al(NO_3)_3$

226. Arrange the following aqueous solutions in order of <u>increasing</u> boiling point.
   a) 0.075 M glucose
   b) 0.075 M LiBr
   c) 0.030 M $Zn(NO_3)_2$
   d) 0.030 M NaI

227. Which aqueous solution would have the lowest boiling point, highest vapor pressure, lowest freezing point?
   a) 1.50 m NaCl   b) 2.00 m CH_3OH c) 0.60 m Na_3PO_4 d) 0.62 m MgF_2

228. Calculate the freezing point of a solution prepared by dissolving 15 g Na_3PO_4 in 135 mL of water. K_f = 1.86 °C /m; density of water = 1.00 g/mL.

229. When 3.70 g of a substance known to be a nonelectrolyte is dissolved in 50.0 g of benzene (C_6H_6), the solution freezes at 1.20 °C. If the normal melting point of benzene is 5.50 °C and K_f for benzene is 5.12 °C/m, what is the molecular weight of the solute?

230. Carbon disulfide, CS_2, boils at 46.30 °C and has a density of 1.261 g mL^{-1}. When 0.250 mol of a non-dissociating solute is dissolved in 400.0 mL of CS_2, the solution boils at 47.46 °C. What is the molal boiling point elevation constant for CS_2? Remember that $\Delta T_b = K_b m$.

231. Calculate the molarity of a solution made by dissolving 10.2 g of sodium phosphate in enough water to form 650 mL of solution.

232. How many mL of 2.00 M NaCl would be needed to prepare 5.00 L of 0.200 M NaCl?

233. Assume 6.73 g of Na_2CO_3 is dissolved in enough water to make 250 mL of solution.
   a) What is the molarity of the sodium carbonate?
   b) What are the concentrations (molarities) of Na^+ and CO_3^{2-} ions?

234. What mass of NaOH is required to precipitate all the copper(II) ions from 100.0 mL of 0.100 M copper(II) nitrate solution?

235. A solution is prepared by adding 25 g NaCl and brought up to a volume of 500 mL in a volumetric flask using DI water (1.0 g = 1.0 mL). Calculate the molarity, molality, ppm, ppb, mass fraction, and weight percent.

236. A solution was prepared by dissolving 15.0 g NH_3(17 g/mol) in enough water to make 250 mL of solution. If the density of the resulting solution is 0.974 g/mL, what is the molarity of ammonia in the solution?

237. A 32.0% by weight solution of propanol, $CH_3CH_2CH_2OH$, in water has a density at 20 °C of 0.945 g mL^{-1}. What are the molarity and molality of the solution?

238. Consider the table below for the reaction A + B $\longrightarrow$ C. What is the average rate between 10 s and 40 s?

Time (s)	[A] (mol/L)
0.0	0.124
10.0	0.110
20.0	0.088
30.0	0.073
40.0	0.054

239. Use the information below to determine the **overall** order of the reaction A + B ⟶ products.

Experiment Number	[A] (*M*)	[B] (*M*)	Initial Rate (*M*/s)
1	0.273	0.763	2.83
2	0.273	1.526	2.83
3	1.365	0.763	70.75

240. Consider the following kinetic data for a **first** order reaction, A ⟶ B. a) What is the average rate of formation of product B between 0 and 9 seconds? b) What is the half-life (in seconds) for the reaction? c) What is the value of the reaction rate constant, k, for the reaction?

Time (seconds)	[A] (*M*)
0	1.22
3	0.86
6	0.61
9	0.43
12	0.31
15	0.22
18	0.15

241. Consider the hypothetical aqueous reaction $A_{(aq)}$ ⟶ $B_{(aq)}$. The following table shows the concentration of A with respect to time. Determine the average rate of the reaction for each 10 minute time interval. Explain the differences in the three values you calculated.

Time (min)	0	10	20	30
Molarity of A	0.065	0.051	0.042	0.036

242. Experiment data are listed below for the hypothetical reaction A ⟶ 2B

Time (s)	[A] (mol/L)
0.00	1.000
10.00	0.833
20.00	0.714
30.00	0.625
40.00	0.555

a) Plot these data, connect the points with a smooth line, and calculate the rate of change of [A] for each 10-second interval from 0 to 40 seconds.

b) Why does the rate of change decrease from one time interval to the next?

c) How is the rate of change of [B] related to the rate of change of [A] in the same time interval?

d) Calculate the rate of change of [B] for the time interval from 10 to 20 seconds.

243. A first order reaction has a rate constant of 0.075 s^{-1}. What is the concentration of the reactant after 15 seconds if the initial concentration is 0.056 M?

244. The catalyzed decomposition of H_2O_2 is a first order reaction with a reaction rate constant of 1.5×10^4 s^{-1} at a certain temperature. If the initial concentration of H_2O_2 is 0.026 M, how long will it take for the concentration of H_2O_2 to decrease to 0.0022 M?

245. If a reaction has the experimental rate expression **rate** = **k[A]²**, explain what happens to the rate a) when the concentration of A is tripled, and b) when the concentration of A is halved.

246. A reaction obeys the following rate law: rate = k[A][B]². a) What will be the affect on the rate constant, k, if the concentration of A is doubled? b) How will the rate of this reaction be affected if the concentration of B is tripled?

247. Which of the following reactions appear to involve a catalyst? In those cases where a catalyst is present, tell whether the catalyst is homogeneous or heterogeneous.

a) $CH_3CO_2CH_{3(aq)} + H_2O_{(l)} + H_3O^+_{(aq)} \longrightarrow CH_3CO_2H_{(aq)} + CH_3OH_{(aq)} + H_3O^+_{(aq)}$

b) $2\ H_{2(g)} + O_{2(g)} + Pt_{(s)} \longrightarrow 2\ H_2O_{(g)} + Pt_{(s)}$

c) $NH_{3(aq)} + CH_3Cl_{(aq)} + H_2O_{(l)} \longrightarrow Cl^-_{(aq)} + NH_4^-_{(aq)} + CH_3OH_{(aq)}$

248. For the following reaction, the rate expressed in terms of SO_2 is

$$2SO_{2(g)} + O_{2(g)} \longrightarrow 2SO_{3(g)}$$

$$\text{rate} = -\Delta[SO_2]/\Delta t$$

What is the rate of the reaction in terms of O_2?

249. For the reaction mechanism below, predict the rate law expression, identify the intermediates, catalysts and determine the overall reaction.

A + B + Z $\longrightarrow$ 2C + D (fast)

2C $\longrightarrow$ E + F (fast)

E + G $\longrightarrow$ H + Z (slow)

250. In the reaction mechanism below: identify the intermediates, overall reaction and rate law expression.

A + B $\longrightarrow$ C + D   Fast

M + E $\longrightarrow$ F + G   Slow

F + C $\longrightarrow$ H   Fast

251. Draw an energy profile diagram for an endothermic reaction. Show on the diagram the ΔH of the reaction and the activation energy for the forward and reverse reaction. On the basis of the diagram you drew, is the reaction product or reactant favored.

252. A catalyst changes

i. the forward activation energy $E_a(f)$.

ii. the reverse activation energy $E_a(r)$.

iii. the enthalpy change ΔH.

iv. the free energy change ΔG.

a) i & ii   b) iii & iv   c) i & iv   d) ii & iii

253. Answer the following questions for the energy diagram pictured below
    a) What is the activation energy in the forward direction for the reaction A + B ⟶ C + D?
    b) Is this an endothermic or exothermic reaction?
    c) What is the overall energy change for the reaction in the forward direction?

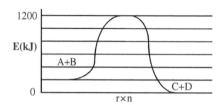

    d) Based on the energy diagram what is the overall order of the reaction?
        i) first in order in A and B
        ii) second order in A first order in B
        iii) first order in A second order in B
        iv) not enough information to determine reaction order

254. For the hypothetical reaction A + B ⟶ C + D the activation energy is 32 kJ/mol. For the reverse reaction (C + D ⟶ A + B), the activation energy is 58 kJ/mol. Is the forward reaction A + B ⟶ C + D exothermic or endothermic?

255. Which one of the following statements about a chemical reaction at equilibrium is correct?
    a) The concentrations of reactants and products are equal.
    b) The rates of the forward and reverse reactions are equal.
    c) No more reactants are converted to products.
    d) The concentrations of reactants and products change with time.

256. Write the equilibrium expression for each equation shown below:
    a) $2\ H_2O_{2(g)} \rightleftharpoons 2\ H_2O_{(g)} + O_{2(g)}$
    b) $SiO_{2(s)} + 3\ C_{(s)} \rightleftharpoons SiC_{(s)} + 2\ CO_{(g)}$

257. Consider the following graph of the first-order reaction of A reacting to produce B. a) Circle the portion of the graph that represents the reaction at equilibrium. b) What is the value of the equilibrium constant (K) for this reaction?

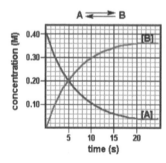

258. a) Write the equilibrium expressions for the following reactions. b) Indicate whether each reaction is homogeneous or heterogeneous.

$N_{2(g)} + O_{2(g)} \rightleftharpoons 2\ NO_{(g)}$ homogeneous or heterogeneous

$Ti_{(s)} + 2\ Cl_{2(g)} \rightleftharpoons TiCl_{4(l)}$ homogeneous or heterogeneous

$4\ HCl_{(g)} + O_{2(g)} \rightleftharpoons 2\ H_2O_{(l)} + 2\ Cl_{2(g)}$ homogeneous or heterogeneous

259. Which diagram pictured below best represents an equilibrium mixture for this reaction if the equilibrium constant is 4.0?

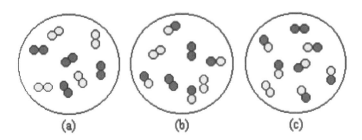

$$A_{2(g)} + B_{2(g)} \rightleftharpoons 2 AB_{(g)}$$

(a)   (b)   (c)

260. Hydrogen gas and iodine gas react via the equilibrium

$$H_{2(g)} + I_{2(g)} \rightleftharpoons 2 HI_{(g)} \qquad K = 7.6 \times 10^3 \text{ (at 727 °C)}$$

If 0.050 mol of HI is placed in a 1.0 L flask at 727 °C, what are the equilibrium concentrations of HI, I_2, and H_2?

261. Hydrogen, bromine, and HBr in the gas phase are in equilibrium in a container of fixed volume.

$$H_{2(g)} + Br_{2(g)} \rightleftharpoons 2 HBr_{(g)} \qquad \Delta H° = -103.7 \text{ kJ}$$

How will each of the following changes affect the <u>equilibrium</u> and the value of the <u>equilibrium constant</u>?
a) Some $H_{2(g)}$ is added to the container.
b) The pressure of $HBr_{(g)}$ is increased.
c) The temperature of this system is increased.

262. Predict the effect on this reaction if the following changes are made.

$$Fe(OH)_{3(s)} + H_2O_{(l)} \rightleftharpoons Fe^{3+}_{(aq)} + 3 OH^-_{(aq)}$$

a) $Fe(NO_3)_{3(aq)}$ is added
b) $NaOH_{(aq)}$ is added
c) $Fe(OH)_{3(s)}$ is added
d) $HCl_{(aq)}$ is added

263. For the reaction $2A_{(g)} + B_{(g)} \rightleftharpoons 2C_{(g)}$ at equilibrium, what is the effect on the equilibrium of:
a) Adding $D_{(g)}$ to the reaction vessel
b) Adding $C_{(g)}$ to the reaction vessel
c) Decreasing the reaction vessel volume
d) Removing $A_{(g)}$ from the reaction vessel

264. A mixture of 0.6816 mol of H_2, 0.4400 mol of Br_2, and 2.268 mol of HBr is in a 2.00 L flask at 700 K. They react according to the following equation:

$$H_{2(g)} + Br_{2(g)} \rightleftharpoons 2 HBr_{(g)}$$

The equilibrium constant for this reaction is 58.4 at 700 K. Is the reaction at equilibrium? If it is not at equilibrium, which way will it shift to attain equilibrium?

265. At 400 K, for the reaction $Br_{2(g)} + Cl_{2(g)} \rightleftharpoons 2 BrCl_{(g)}$ the K = 7.0. If 0.300 mol of Br_2 and 0.300 mol of Cl_2 are introduced into a 1.00 L container at 400 K, what will be the equilibrium concentrations of Br_2, Cl_2, and BrCl?

266. Consider the following reaction

$$NH_4Cl_{(s)} \rightleftharpoons NH_{3(g)} + HCl_{(g)} \qquad \Delta H = 176 \text{ kJ}$$
$$K_{eq} = 1.1 \times 10^{-4} \text{ at } 1200 \text{ K}$$

If the above system is at equilibrium, adding some $NH_{3(g)}$ will

a) shift the reaction to right.
b) shift the reaction to the left.
c) decrease the equilibrium constant.
d) increase the equilibrium constant.
e) decrease the amount of HCl.
f) increase the amount of NH_4Cl.
g) have no effect.

267. $PbCl_{2(s)}$ is a slightly soluble salt, $K_{sp} = 1.7 \times 10^{-5}$. Calculate the Cl^- concentration for a saturated solution of $PbCl_{2(s)}$.

268. The K_{sp} for $Fe(OH)_{3(s)}$ is 4.0×10^{-38}. Calculate the molar solubility of $Fe(OH)_{3(s)}$

269. Calculate the molar solubility of $Cu(OH)_{2(s)}$ given its K_{sp} is 4.8×10^{-20}.
The molar solubility of $Cu(OH)_{2(s)}$ in 0.1 M NaOH would be (explain):
a) Greater than in DI water
b) Less than in DI water
c) The same as in DI water

270. The molar solubility of $Cu(OH)_{2(s)}$ in 0.1 M HCl would be (explain):
a) Greater than in DI water
b) Less than in DI water
c) The same as in DI water

271. What is the solubility of $PbBr_{2(s)}$ in a 0.15 M solution of $Pb(NO_3)_2$? ($K_{sp} = 6.3 \times 10^{-6}$)

272. The K_{sp} of $Fe(OH)_{3(s)}$ is 4.0×10^{-38}. Calculate the molar solubility of $Fe(OH)_3$ in a solution with a pH of 13.6.

273. Simple acids like formic acid, HCOOH, and acetic acid, CH_3COOH, are very soluble in water. However, larger more complex acids such as stearic acid, $CH_3(CH_2)_{16}COOH$ and palmitic acid, $CH_3(CH_2)_{14}COOH$ are insoluble in water. (Stearic acid and palmitic acid belong to a class of compounds known as fatty acids.) Explain these differences in solubility.

274. Calculate the hydroxide concentration of a 0.15 M solution of HCl.

275. What is the pH of a 0.0012 M solution of HCl? What is the OH^- concentration? What is the pOH?

276. The pH of a solution of NaOH is 10.23. What is the OH^- concentration? What is the pOH? What is the H^+ concentration?

277. Convert 1.7×10^{-5} M hydronium ion concentration to pH, hydroxide concentration and pOH.

278. Convert a pOH of 3.42 to hydrogen ion concentration, pH and hydroxide ion concentration.

279. Write a reaction of HF (a weak acid) with water. Circle the substance in the reaction that is acting as a base. Underline the conjugate base and put a box around the conjugate acid. Write the K_a expression for HF.

280. Write a reaction of NH_3 (a weak base) with water. Circle the substance in the reaction that is acting as an acid. Underline the conjugate base and put a box around the conjugate acid. Write the K_b expression for NH_3.

281. Write an equation to describe the proton transfer than occurs when each of the following acids or bases is added to water. The first one is done as an example.
a) $HCN + H_2O \rightleftharpoons H_3O^+ + CN^-$
b) F^- (base)
c) HCO_3^- (base)
d) $H_2PO_4^-$ (acid)

282. Based on formulas alone, which member of each pair is the stronger acid?
    a) H_2CO_3 or H_2SO_4
    b) HNO_3 or HNO_2
    c) H_2SO_3 or H_2SO_4

283. Which are conjugate acid-base pairs?
    a) NH_2^- and NH_4^+ b) OH^- and O^{2-}
    c) H_3O^+ and OH^- d) NH_3 and NH_2^-

284. Calculate the hydroxide concentration of a 0.15 M solution of acetic acid. $K_a = 1.8 \times 10^{-5}$.

285. The K_a for hypochlorous acid is 3.0×10^{-8} and the K_a for chlorous acid is 1.1×10^{-2}.
    a) Which is the stronger acid, $HClO$ or $HClO_2$?
    b) Which is the stronger base, ClO^- or ClO_2^-?
    c) Calculate the pH of a 0.25 M solution of $HClO$.

286. Circle the member of each pair that is the stronger acid. Explain your answers based on the structure of the acid.
    a) HNO_3 or HNO_2 b) HCl or H_2S c) CCl_3COOH or CH_3COOH

287. What is the pH of a 0.10 M aqueous solution of sodium acetate ($NaC_2H_3O_2$) if $K_a = 1.8 \times 10^{-5}$ for acetic acid?

288. Predict whether aqueous solutions of the following salts will be acidic, basic, or neutral. If the solution would be acidic or basic, circle the ion that causes the pH to change.
    a) NH_4Br b) Na_2CO_3 c) $KClO_4$

289. For which of the following substances would the solubility be greater at pH=2 than at pH=7?
    a) $Cu(OH)_2$
    b) CuS
    c) $CuSO_4$
    d) $CuCO_3$

290. Circle the member of each pair which produces the more acidic aqueous solution. Explain your answers.
    a) K^+ or Cu^{2+} b) Fe^{2+} or Fe^{3+} c) Al^{3+} or Ga^{3+}

291. Predict the effect on this reaction at equilibrium if the following changes are made.

$$HCN_{(aq)} + H_2O_{(l)} \rightleftharpoons H_3O^+_{(aq)} + CN^-_{(aq)}$$

    a) $HCl_{(aq)}$ is added
    b) $NaOH_{(aq)}$ is added
    c) $NaCN_{(aq)}$ is added

292. Which of the following pairs could be used to make a buffer solution? Explain.
    a) HCl, KCl b) HNO_2, $NaNO_2$ c) HNO_3, KNO_3
    d) NH_3, NH_4Cl e) HF, $NaOH$

293. The K_a for hydrofluoric acid is 7.2×10^{-4}. Calculate the pH of a 0.35 M aqueous solution of HF.

294. A 0.015 M solution of hydrocyanic acid has a pH of 5.61. What is the acid ionization constant (K_a) for HCN?

295. Consider 1.0 L of an aqueous buffer solution that is 0.10 M acetic acid (HOAc) and 0.13 M sodium acetate (NaOAc). (The K_a of acetic acid is 1.8×10^{-5}.)
    a) What is the pH of this buffer solution?
    b) What is the pH of the buffer solution after the addition of 0.010 mol of HCl?

296. Calculate the pH of a buffer prepared by dissolving 0.25 moles of NaF and 0.10 M of HF in 1.0 L of water. What is the effect of adding 0.08 moles of HCl to this solution? (F^-, $K_b = 1.4 \times 10^{-11}$)

297. a) Calculate the pH of a solution prepared by dissolving 0.75 moles of NH_3 and 0.25 moles of NH_4Cl in water sufficient to yield 1.00 L of solution. The K_b for ammonia is 1.8×10^{-4}. b) Now calculate the pH of the solution after adding 0.075 moles of NaOH.

298. The pH of a buffer solution containing equimolar amounts of a weak acid and its conjugate base is 4.56. What is the K_a of the acid?

299. What volume of 0.105 M NaOH is required to titrate each of the following solutions to the equivalence point? In each case indicate whether the equivalence point is acidic, basic, or neutral.
    a) 20 mL of 0.075 M HNO_3
    b) 15 mL of 0.165 M formic acid (HCO_2H)

300. The equivalence point for an acid base titration curve is at pH 8.20. What type of acid/base titration is this?
    a) A strong acid titrated with a strong base.
    b) A weak acid titrated with a strong base.
    c) A weak base titrated with a strong acid.
    d) A strong acid titrated with a weak base.

301. Sketch the titration curve that would result if 25 mL of a 0.100 M solution of ammonia (NH_3, $K_b = 1.8 \times 10^{-5}$) were titrated with 0.125 M HCl.

$$NH_{3(aq)} + H_3O^+_{(aq)} \longrightarrow NH_4^+_{(aq)} + H_2O_{(l)}$$

Be sure to point out the pK_a and the equivalence point for this titration.

302. For each of the following chemical equations, write the complete and balanced oxidation and reduction **half-reactions**.
    a) $NO_2^-_{(aq)} + Cr_2O_7^{2-}_{(aq)} \longrightarrow Cr^{3+}_{(aq)} + NO_3^-_{(aq)}$ (acidic solution)
    Oxd ½–rxn:
    Rdn ½–rxn:

    b) $MnO_4^-_{(aq)} + Br^-_{(aq)} \longrightarrow MnO_{2(s)} + BrO_3^-_{(aq)}$ (basic solution)
    Oxd ½–rxn:
    Rdn ½–rxn:

303. Identify the **oxidizing** and **reducing agents** in each of the following reactions.
    a) $I^-_{(aq)} + Cr_2O_7^{2-}_{(aq)} \longrightarrow Cr^{3+}_{(aq)} + IO_3^-_{(aq)}$ (acidic solution)
    b) $I_{2(s)} + OCl^-_{(aq)} \longrightarrow Cl^-_{(aq)} + IO_3^-_{(aq)}$ (acidic solution)

304. Determine the emf ($E°$) under standard conditions of a voltaic cell based on the following chemical reaction.

$$Ce^{4+}_{(aq)} + Ni_{(s)} \longrightarrow Ni^{2+}_{(aq)} + Ce^{3+}_{(aq)}$$

305. In each of the following balanced equations, identify those elements that undergo a change in oxidation number. If an element in a reactant undergoes a change in oxidation number, indicate whether it is being oxidized or reduced.
    a) $4\,Fe_{(s)} + 3\,O_{2(g)} \longrightarrow 2\,Fe_2O_{3(s)}$
    b) $BaCl_{2(aq)} + 2\,NaOH_{(aq)} \longrightarrow Ba(OH)_{2(s)} + 2\,NaCl_{(aq)}$
    c) $I_2O_{5(s)} + 5\,CO_{(g)} \longrightarrow I_{2(s)} + 5\,CO_{2(g)}$